D0700200

Astronomers' Universe

Other titles in this Series

Origins: How the Planets, Stars, Galaxies,
and the Universe Began
Steve Eales

Calibrating the Cosmos: How Cosmology Explains our
Big Bang Universe
Frank Levin

The Future of the Universe
A. J. Meadows

It's Only Rocket Science: An Introduction to Space
Enthusiasts (forthcoming)
Lucy Rogers

Vladimir Rubtsov

The Tunguska Mystery

Vladimir Rubtsov
P. O. Box 4542
Kharkov 61022, Ukraine
tunguskamystery@gmail.com

ISBN 978-0-387-76573-0 e-ISBN 978-0-387-76574-7
DOI 10.1007/978-0-387-76574-7
Springer Dordrecht Heidelberg London New York

Library of Congress Control Number: 2009931092

Printed on acid-free paper

Springer is part of Springer Science+Business Media (www.springer.com)

Contents

Acknowledgments

First of all, I would like to thank the editor of this book, Edward Ashpole. Without his participation, this book would never have been written – at least, not in the English language. Apart from the essential linguistic comments and corrections, Edward's critical eye and acute insight have been very helpful and inspiring. It was a real pleasure to work together during these months, preparing this exposition on one of the most enigmatic events that have ever occurred on our planet.

Our editor at Springer, Maury Solomon, should be praised for her keen interest in the subject and patient waiting for the final version of the manuscript to be produced.

From personal experience I can say that the enigma of Tunguska is utterly compelling. Hardly any of those who became well acquainted with it (especially in his or her younger years) has been able to put it aside. Physicists and engineers, geneticists and geologists, science amateurs and professionals – all of them could find in this field something to attract them. Having once set foot on the Tunguska road, these people still persistently continue their quest for truth despite so many different obstacles. Even dissenting between themselves, they comprise a research community united by the same goal – and by the same values. I wish to express my sincere thanks to all friends and colleagues with whom we have been traveling on this road – first of all to Dr. Victor Zhuravlev, whose advice and encouragement have always been so helpful. Victor is remarkable for his wonderful ability to wed intellectual bravery with strict logic – a true man of science.

Also, I am deeply indebted for help, criticism, and useful discussions to Alexander Beletsky, Boris Bidyukov, Rod Brock, Victor Chernikov, Robert Dehon, Hilary Evans, Mikhail Gelfand, Dr. Lev Gindilis, Robert Gray, Dr. Viacheslav Gusiakov, Dr. Stanislav Kriviakov, Dr. Pyotr Kutniuk, Dr. Yuly Platov, Dr. Gennady Plekhanov, Vitaly Romeyko, Dr. Vladimir Shaydurov, Dr. Mikhail

Shevchenko, Dr. Vitaly Stepanok, and Dr. Kazuo Tanaka. And it is with heartfelt gratitude that I wish to remember here those friends who are no more: Alexander Kazantsev, Alexey Zolotov, Nikolay Vasilyev, Felix Zigel, and Sokrat Golenetsky. They did not live to see the Tunguska problem finally solved, but it is due to them that today we can correctly judge its scope, complexity, and significance. And I have no doubt that their works will become a cornerstone for the future final solution of this problem.

Author's Note

This book is different from previously published books on the greatest explosion in recorded history in that it offers a truly interdisciplinary approach to the subject. Unfortunately, many theorists who try to solve this enigma are aware of only two facts: in 1908 something flew over Siberia, and this "something" exploded. Actually there is far more to this story. I personally researched this subject for 35 years and this book presents the wealth of information collected in Russia during the past 100 years.

Theories that attempt to explain what happened at Tunguska in 1908 must use all the facts established by hundreds of investigators (scientists and their assistants) on numerous expeditions since the 1920s. Some theories have come close to doing so, although none has fully satisfied the available data, much of which have only been recorded in Russian. Readers will soon see that this subject is much more complex than was once thought, and that the interdisciplinary approach seems to offer the only way of knowing what actually hit Earth with such force in 1908.

1. The Enigma of Tunguska

The summer of 1908 witnessed the arrival of an unknown space body and an explosion over the Tunguska forest in Central Siberia that could have flattened any major city on Earth. The Tunguska explosion has been publicized in the popular press and scientific journals for decades, yet both the general public and the science community still seem unaware of the complicated details of this event. The key publications are in Russian, so language has been a barrier to understanding the evidence of what took place. Most people think that the Tunguska event was explained long ago by scientists who study meteorites or that the incident remains unimportant as far as science is concerned. Neither of these assumptions is anywhere near the truth. And what has been discovered in recent decades raises startlingly complex questions.

Strange as it may seem, the Tunguska event did not begin with a big bang. Scientists recorded the occurrence of some unusual phenomena starting on June 27, 1908.[1] That was three days before the devastating explosion. Some specialists even suppose that these phenomena started as early as June 23 or June 21, but for these dates the supporting evidence is scarce. Optical anomalies in the atmosphere (strange silvery clouds, brilliant twilights, and intense solar halos) were observed in western Europe, the European part of Russia, and western Siberia. The farthest western point from where these anomalies were recorded seems to have been Bristol in England. William F. Denning (1848–1931), a noted British specialist in meteors, wrote in *Nature* in 1908 that on the night of June 30, the firmament over Bristol was unusually light and few stars could be seen.[2] The whole northern part of the sky was red-colored, while the eastern part looked green.

The anomalies increased in intensity during the three days prior to the sunny morning of June 30, 1908, when a fiery body flew over central Siberia, moving in a northerly direction. It was seen from many settlements in the region, its flight being

V. Rubtsov, *The Tunguska Mystery*, Astronomers' Universe,
DOI 10.1007/978-0-387-76574-7_1, © Springer Science+Business Media, LLC 2009

accompanied by thunderous sounds. Because this region is remote and sparsely populated, the systematic gathering of eyewitness reports was only begun in the 1920s. However, we now have some 500 written accounts that contain more or less detailed descriptions of the flying body, its shape being mostly described as roundish, spherical, or cylindrical, and its color as red, yellow, or white. What is important is that no one reported a smoky trail, which is typical for large iron meteorites traveling through the atmosphere, although many witnesses saw vivid iridescent bands, like a rainbow, behind the space body.

When flying at 0 h 14 min GMT over the so-called Southern swamp, a small morass not far from the Podkamennaya Tunguska River (see Figure 1.1), the body exploded, releasing the TNT equivalent of 40 to 50 megatons (Mt) of explosive. That is equivalent to 3,000 atomic bombs of the kind dropped on Hiroshima in 1945.[3] There was a brilliant flash and a devastating blast. Had this occurred over London or New York an entire city would have been destroyed. Was it a meteorite? Unlikely. Was it a comet? Or was it something else, perhaps something that only advanced physics could explain?

FIGURE 1.1. The Southern swamp, where the Tunguska meteorite exploded. View from a helicopter (*Photo by* Vladimir Rubtsov).

In 1927, Semyon Semyonov, a local farmer who then lived in the small trading station of Vanavara, 70 km south-southeast from the epicenter of the explosion, the closest settlement to the catastrophe, recalled his experience: "I sat on the steps of my house facing north. Suddenly the sky in the north split apart, and there appeared a fire that spread over the whole northern part of the firmament. At this moment I felt intense heat, as if my shirt had caught fire. I wished to tear my shirt off and throw it away, but at this moment a powerful blast threw me down from the steps. I fainted, but my wife ran from the house and helped me up. After that we heard a very loud knocking, as if stones were falling from the sky."

The Evenks (or Tungus), the native inhabitants of the region, were also much impressed by what happened. Two Evenk brothers, Chuchancha and Chekaren, were at the moment of the explosion sleeping in their *chum* (a tent of skin or bark) on the bank of the Avarkitta River some 30 km to the south-southeast from the epicenter of the explosion. They had returned just before sunrise from a long trip to the Dilyushma River. Suddenly the brothers were woken by tremors and the noise of the wind. "Both of us were very frightened," Chuchancha in 1926 told the anthropologist Innokenty Suslov: "We began to call our father, mother, and third brother, but nobody replied. We heard a loud noise from outside the *chum*. Trees were falling. Chekaren and me got out of our sleeping bags and were going to get out of the *chum*, but suddenly there was a great clap of thunder. The ground trembled, and a strong wind hit our *chum* and threw it down. The *elliun* (the skins covering a *chum*) rode up, and what I saw was terrible. Trees were falling down, their pine needles burning. Branches and moss on the ground were burning as well. Suddenly a bright light like a second Sun appeared above the mountain where the trees had fallen. At the same moment a strong *agdyllian* (thunder) crashed. The morning was sunny with no clouds. The Sun shone as always, and now there was a second Sun. Chekaren and I crawled out from under the *chum*. After that we saw another flash of light while thunder crashed overhead followed by a gust of wind that knocked us down. Then Chekaren cried out: 'Look up!' and stretched his hand upward. I looked and saw new lightning and heard more thunder."

The Tunguska explosion was heard more than 800 km from the epicenter, and within 200 km some windows facing north were broken. The seismic wave was recorded in Russia at Irkutsk, Tashkent, Tbilisi, and in Germany at Jena. The shock wave leveled more than 2,100 km^2 of the forest. Over an area of 200 km^2 vegetation was burnt by the flash that produced a major forest fire. Minutes after the explosion a magnetic storm began, similar to the geomagnetic disturbances following nuclear explosions in the atmosphere. This was detected by the Magnetographic and Meteorological Observatory in Irkutsk. The storm lasted 5 hours. By the dawn of July 1 the strange lighting effects in the skies, which had started four days earlier, reached their peak and had begun to fade, although aftereffects persisted till late July.

Even this brief introduction to the Tunguska phenomenon shows its puzzling aspects. So, the lack of any serious reaction to it by scientists at the time seems more than odd. Some scientific journals did discuss the atmospheric anomalies, but the attention this whole subject received hardly matched the extraordinary event that had leveled some 30 million trees and devastated part of Siberia. Some local Siberian newspapers did, however, publish eyewitness accounts that led to journalists writing that a huge meteorite had hit the taiga. The very first but partly fictitious article entitled "A Visitor from Heavenly Space" appeared on July 12 in the newspaper *Sibirskaya Zhizn* (*Siberian Life*) that was published in the city of Tomsk. The reporter Alexander Adrianov wrote: "A terrible rumble and a deafening thud were heard 40 km away. A train that was approaching the station of Filimonovo was stopped by its driver, and the passengers rushed to view the cosmic visitor that had fallen from the sky. But it was impossible to examine the burning hot meteorite in any detail. Later, when the meteorite cooled, it was trenched around and examined by many people from Filimonovo..." Almost everything in this story is due to the imagination of the reporter. But this article was later seen by meteorite specialist Leonid Kulik, who was to play a major role in the story of the Tunguska event, and it motivated him to search for what was initially named the "Filimonovo meteorite."

The second newspaper article was published on July 15, 1908, in the newspaper *Sibir* (*Siberia*), and its author was more accurate: "On June 30, soon after 8 o'clock, there occurred in our region an unusual phenomenon of nature. In the village of Nizhne-Karelinskoye [some 450 km from the epicenter] peasants saw in the north-west, high above the horizon, a blindingly bright body of bluish-white color that was flying above for about 10 min. The body looked like a tube. The sky was cloudless, but one could see a small dark cloud in the same direction where the luminous body was observed, low above the horizon. Having approached the forest the luminous body became blurred. There was an enormous mass of black smoke and a loud knocking, but not of thunder. The buildings were trembling and a fire of indefinite shape gushed out from the small dark cloud. All the village inhabitants ran from their houses in terror. Women were crying and everyone thought Armageddon had arrived."

In 1921, an expedition of the Russian Academy of Sciences, led by the just-mentioned Leonid Kulik, visited central Siberia to gather information about meteorites in general, and during this expedition Kulik collected new eyewitness reports of the Tunguska event. There seemed to be no question that it had been a huge meteorite, most likely of iron. A few years later, in 1927, Kulik discovered the huge area of leveled forest that marked the place of the Tunguska "meteorite" fall. Subsequently, several well-equipped expeditions were sent to the site, and Kulik continued to explore the area until World War II.

However, even the expedition of 1927 made the surprising discovery that at the actual epicenter of the explosion the trees were still standing and that there was no sign of a large meteorite crater. It seems strange now that at the time no real significance was attached to this. There was just a little shift from the idea of a single meteorite to a shower of meteorites from a body that broke up due to air resistance above Earth's surface. The forest was therefore supposed to have been flattened by the ballistic shock wave from the disintegrating body – by the air compressed by the body in flight. At the time, Leonid Kulik mistook what are called thermokarst holes for numerous meteorite craters. (Thermokarst holes are shallow depressions caused by selective thawing of ground ice or permafrost.) However, Kulik should perhaps not be faulted for this mistake. He was a specialist on meteorites and therefore looked for evidence of a meteorite – not for something else.

Nevertheless, as time passed, some scientists felt that the meteorite hypothesis was flawed. In spite of extensive searches for remnants of the meteorite, none were found. So, in the early 1930s, British astronomer and meteorologist Francis Whipple suggested that the Tunguska space body had been the core of a small comet. The geochemist Vladimir Vernadsky, who was then famous both in the Soviet Union and in Europe, favored a lump of cosmic matter (something like a compact cloud of cosmic dust), while astronomer Igor Astapovich assumed that a meteorite body had ricocheted off a lower layer in the atmosphere. But it was the Russian engineer and science fiction writer Alexander Kazantsev who in 1945 suggested an even stranger explanation for the Tunguska event. He enraged the science community by suggesting that the data then available testified to the possibility of an extraterrestrial spaceship meeting disaster in the final stage of its voyage. At the time he said he had been much impressed by the similarities in the description of the Tunguska event and those describing the nuclear explosion over Hiroshima.

As one can imagine, the meteorite specialists "were not amused." They at once objected to such a fantastic idea, and in 1951, a team of the most distinguished Soviet astronomers expressed their opinion in the popular science journal *Nauka i Zhizn* (*Science and Life*). "There is," they said, "no question that immediately after the meteorite fall a crater-like depression formed where now the Southern swamp exists. It was relatively small and soon became inundated with water. In subsequent years it was covered by silt and moss, filled with peat hummocks and partly overgrown with bushes. The dead trees standing upright can be seen not at the center of the catastrophe, but on the hillsides which surround the hollow."

This was what the then leading Soviet astronomers accepted, being absolutely certain that the Tunguska event had been due to a normal stone or iron meteorite. Consequently, they rejected even the most obvious facts, such as the location of the standing trees at the epicenter of the devastation. And they were equally certain that there had to be a crater at Tunguska. However, the first postwar Tunguska expedition, organized in 1958 by the Committee on Meteorites of the USSR Academy of Sciences, made everyone involved agree that the Tunguska space body had exploded in the air and therefore could hardly have been a normal meteorite. At least that much was accepted.

From then on the number of anomalies discovered at the site of the Tunguska explosion began to grow very fast. And the hypotheses that the Tunguska space body was a meteorite or the core of a small comet met with considerable difficulties. Thus in 1962, the Committee on Meteorites turned the problem over to the Commission on Meteorites and Cosmic Dust of the Siberian Branch of the USSR Academy of Sciences. The problem of the Tunguska phenomenon was exiled to the place of its birth.

In 1958, the so-called Independent Tunguska Exploration Group was established under the leadership of young Siberian scientists Gennady Plekhanov and Nikolay Vasilyev. This group became responsible for the ensuing Tunguska studies and initially consisted of a dozen specialists, mainly physicists and mathematicians. Actually, this organization was conceived for the purpose of settling only one persistent question that by then had gained an embarrassing prominence in the Soviet Union. It was whether or not the Tunguska space body had been an extraterrestrial spaceship. But this led to the realization that the problem of the Tunguska event would require a lot more research, involving high-level specialists applying the latest know-how and technology. Consequently, within a few years, the "core" of this organization would consist of 50 scientists, while a 100 specialists would take part in fieldwork each year with an amazing 1,000 researchers from various scientific institutions all over the Soviet Union collecting and analyzing relevant materials.

In 1959, geophysicist Alexey Zolotov, a specialist in using nuclear physics to examine geological deposits, suggested ways of testing the main aspects of the spaceship hypothesis. He asked whether it was an explosion in the usual sense of this word that devastated the taiga of the Tunguska or was it a ballistic shock wave from a moving space body? If it was an explosion, was it a nuclear explosion or not? Alexander Kazantsev, the science fiction writer, believed it was nuclear, or something similar, while fully realizing that one could hardly imagine an alien spaceship carrying a nuclear reactor similar to those built in the United States and USSR in the 1940s. Still less could one imagine interstellar travelers having an atomic bomb aboard. Nevertheless, if significant traces of nuclear reactions were discovered in the taiga, the "meteorite model" would have to be reconsidered. Alexey Zolotov did succeed in answering the first question: Yes, it was an explosion and not a ballistic shock wave. In other words, the destruction of the forest was due to the energy of an exploding body, not due to the force of energy produced by such a body's motion through the atmosphere.

That, as we shall see, was very important. But the second question remained unresolved. There were nuclear traces on the site but they were too feeble for any conclusion.

In recent decades the Tunguska event has become a major problem for many scientists who have their own publications and research communities to consider, although scientists in the Russian "meteoritic establishment" are definitely not ready to consider the "spaceship hypothesis." They regard this as a terrible heresy, even though Vasilyev, Zolotov, Plekhanov, and others have examined the hypothesis with rigorous scientific research methods. So from 1946 (when Alexander Kazantsev publicized the Tunguska event by publishing his heretical hypothesis), there have been two groups in the Soviet Union that have led a not-so-peaceful coexistence. The natural explanation versus the artificial explanation has remained the keynote in the whole Tunguska affair during the last 60 years. This situation may surprise scientists in the West, but whatever model of the space body turns out to be correct, this competition between the two camps has at least been very productive. Without this controversy every astronomer would have automatically assumed that an icy core of a comet caused the Tunguska event – and nothing else. Some astronomers might even have been awarded the State Lenin Prize of the USSR for such an epoch-making discovery. This was actually planned in the early 1960s.

After the expedition of 1961, Kirill Florensky (a noted geochemist and head of the academic Tunguska expeditions) asserted categorically that the problem of the Tunguska event had been solved. The space body was indeed a comet. Of course, everyone has the right to proclaim what he or she believes correct, but the spicy detail is that the scientists responsible for this outstanding scientific result were thought worthy of a "State Lenin Prize of the USSR." Being a laureate of this prize carried great weight in Soviet times, but in this case any prospects for serious Tunguska studies would have been closed for years to come. However, Gennady Plekhanov and his friends, not agreeing with "the comet solution," threatened to raise hell in the newspapers, and the establishment meteor specialists had to retreat. There was no further collaboration between the two camps.

In the 1970s, the author of this book worked for several years in the Russian town of Kalinin (now Tver) in the laboratory of Dr. Zolotov. It was a small unit in a big geophysical institute. The scientists there called it the "Laboratory of Anomalous Geophysics." It had only four staff: Alexey Zolotov, Sokrat Golenetsky, Vitaly Stepanok, and myself, a recent graduate of Kharkov Polytechnical Institute. Golenetsky and Stepanok were looking for material and radioactive traces of the Tunguska meteorite, whereas I was mainly engaged in computer processing the collected data. When in Moscow we often met with science fiction writer Alexander Kazantsev and some Siberian Tunguska specialists with whom we discussed the scientific approaches to the Tunguska problem. Subsequently, while working on my dissertation on the scientific searches for extraterrestrial intelligence, I used the Tunguska "natural versus artificial" competition to illustrate the justification of the two approaches to such a problem and the need to investigate both with the same scientific rigor.

In 1992, a group of scientists, scholars, and engineers, living in different countries but equally interested in scientific research on anomalous phenomena of various kinds, established the interdisciplinary Research Institute on Anomalous Phenomena (RIAP). By mutual agreement it was established in the Ukrainian city of Kharkov, and one of the main research topics was – and still is – the Tunguska problem. The Tunguska investigations at RIAP are carried out in collaboration with the Independent Tunguska Exploration Group that still exists as an "invisible college" throughout the territory of the Community of Independent States. Consequently, Russian Tunguska investigators today have a niche in the new, postcommunist socioeconomic order. True, the large and costly expeditions of Soviet times are a thing of the past, but the National Nature Reserve *Tungussky* has been established by the Russian Federal Government, and the area of the explosion is not standing empty. Even tourists from abroad visit the region, mainly in summer, and conferences are organized by scientific institutions in Moscow, Krasnoyarsk, Tomsk, Novosibirsk, and other Russian cities. As for the scientific and popular science publications on this subject, there are, in Russian, hundreds of serious papers and some 50 monographs, all virtually unknown in the West. Although, from time to time, there flashes a spark of interest among Western journalists and TV people – more often than not generated by another flimsy "hypothesis" that has little to do with serious research – the truth of the Tunguska situation is never explained. However, the subject is not unfathomable. At least the problem to be solved can now be well understood, and in this book you will find out about the discoveries made by past investigations as well as about the important questions we have to answer to discover the true nature of the Tunguska catastrophe.

During the twentieth century, the public has often read "The Great Enigma of the Tunguska Meteorite Has Been Solved!" But such statements were premature. Scientific research starts from *seeing a problem*. It is a crucially important stage on the way to real knowledge. With all due respect to Leonid Kulik and his fellow researchers before World War II, their iron meteorite model of the Tunguska space body was based on an inadequate understanding of the problem, so that the hypotheses most seriously considered during the last century may be wrong. However, we do now have the opportunity to solve the problem. For that we need to harness the facts already discovered and build an interdisciplinary picture of the Tunguska event. Of course, some essential bits of empirical information are still needed, and these will have to be gathered from the site. But the amount of data needed will not be very large because the road to a final solution of the Tunguska problem has already been paved by generations of Tunguska researchers.

Notes and References

1. In 1908, the Julian calendar was in use in Russia, but to avoid confusion, all dates in the book are given by the Gregorian calendar.
2. See *Nature*, 1908, Vol. 78, No. 2019, p. 221.
3. The TNT equivalent of the bomb dropped on Hiroshima was 13 kilotons (kt). Dividing 50 Mt (that is 50,000 kt) by 13 kt we obtain 3,846. Even if we limit the Tunguska explosion's TNT equivalent to 40 Mt, the result will be 3,077. But of course, the effect of one super powerful explosion is considerably less devastating than that of a group of less-powerful ones. Three thousand "Tunguska mini-meteorites," each of them exploding with the magnitude of 13 kt, would have flattened a much greater area of the taiga than happened in reality.

2. The Big Bang of More than Regional Significance

Let us look back about a 100 years and imagine that we live at the beginning of the twentieth century. This is the starting point of a scientific and technological revolution that will not only transform the world and the material life of European civilization but also transform science itself. But that revolution is only just beginning. Science is not as rich as it will become, but it is freer. Narrow specialization is not that popular in the scientific community, which still has scholars with encyclopedic knowledge who venture to think about things outside their specialty. And there are plenty of naturalists who are interested in the real world more than in the theoretical schemes that represent it. But the mechanisms of human cognition are already undergoing deep changes: science and technology are forming a conglomerate that will soon alter civilization on this planet.

The Wright Brothers' *Flyer I* has just felt air under its wings while a modest schoolteacher in Russia is already developing the theory of jet propulsion that will take humanity into space. That schoolteacher's name is Konstantin Tsiolkovsky, and his paper "Investigation of outer space with jet devices" is published in 1903 by the Russian journal *Nauchnoye Obozreniye* (*Scientific Review*). Max Planck in 1900 lays the foundation of quantum mechanics on which, 13 years later, Niels Bohr will build the first floor of this great edifice, postulating the conditions needed for the existence of stable orbits for electrons in atomic theory. A decade later, a handful of unbelievably gifted people, including Werner Heisenberg, Louis de Broglie, Erwin Schroedinger, and Max Born, will erect on this foundation the edifice itself: a construction of singular beauty and depth. Albert Einstein in 1905 had created the Special Theory of Relativity, and after 10 years of thought experiments and calculations the General Theory of Relativity.

There was also research to confirm new sensational physical theories, in particular Eddington's observations of a solar eclipse

V. Rubtsov, *The Tunguska Mystery*, Astronomers' Universe,

DOI 10.1007/978-0-387-76574-7_2,

that confirmed Einstein's theory of gravitation. Such advances turned human eyes to the heavens, and the prestige of astronomy, though still a science distant from terrestrial needs, rose swiftly, as did the study of meteorites, an interdisciplinary field combining astronomy, geophysics, and geology. Large collections of meteorites – straight from space – had already been gathered. The once heretical conclusion of German naturalist Peter Pallas and physicist Ernst Chladni that meteorites are genuine rocks from space had by then been fully accepted by the scientific community. So 40 years before the Tunguska explosion, the British scientist Nevil Story-Maskelyne had developed in the 1860s the first classification system for meteorites, putting them into three major classes: aerolites (stones), siderites (irons), and mesosiderites (stony irons).

Nowadays we find nothing odd in the fact that stones can fall from the sky – sometimes very large stones. To be convinced of this, just look at the famous Arizona meteor crater. But at the beginning of the twentieth century, some geologists believed that an explosion of volcanic steam had produced this crater. It was not until 1906 that the mining engineer Daniel Moreau Barringer and the mathematician and physicist Benjamin Chew Tilghman published their hypothesis that this immense hole had been formed when a huge meteorite struck Earth that scientists began to take this subject seriously. But even in 1906 not everyone was ready to believe such a mad idea, and it took some years to prove the hypothesis. Nevertheless, the idea spread that the heavens are not always serene and may even be a source of danger. In 1910, lots of people thought that the gigantic tail of Halley's comet, which was known to contain carbon monoxide and cyanogens, might poison the atmosphere and destroy all life on Earth. Consequently, in this context, news of an enormous flying bolide that exploded over distant Siberia should have attracted serious interest both in the science community and among the general public. But due to an unfortunate concurrence of circumstances nothing of this sort happened – at least not in 1908. Several factors affected the situation, the remoteness of the site of the explosion being one factor but not the main one.

So what should have attracted the attention of the science community to this event? There were four initial sources of information that might have stimulated scholars to start investigations:

(1) The descriptions of optical anomalies in the atmosphere over a great part of Eurasia, which occurred from June 27 to July 2 and especially on the night of June 30–July 1.
(2) Data about the flight of an enormous bolide over central Siberia that was recorded in many newspaper articles containing eyewitness testimonies.
(3) The answers from members of the official net of earthquake observers to special questionnaires sent out by Arkady Voznesensky, Director of the Magnetographic and Meteorological Observatory in Irkutsk.
(4) The data on the explosion of the "meteorite" recorded by instruments at the Magnetographic and Meteorological Observatory (and at other observatories) and correctly interpreted by Voznesensky.

Yet all this did not provoke a shift toward recognizing the existence of a big problem that should be solved. Why did it happen so?

Let us first consider the anomalous atmospheric phenomena that both preceded and followed the Tunguska explosion. This is crucial because these phenomena proved to be the *global* trace of this event. Already in the summer of 1908 a possible connection between the atmospheric phenomena and the impact of a large bolide somewhere was suspected. The Russian astronomer Daniil Svyatsky suggested as much although he was then still unaware of the Tunguska event.[1] Some scientists of the time also knew that these optical anomalies lasted from June 27 to July 2 – and even later.[2] These atmospheric anomalies obviously presented a problem because the arrival of a stone or iron meteorite could not account for them. The terrestrial atmosphere could not "prepare itself" for a visiting meteorite, however large, during several days *before* its actual fall. Having seen similar but weaker phenomena in 1910 – after Earth traversed the tail of Halley's comet – the German astronomer Max Wolf, then Director of the Heidelberg Observatory, suggested that the atmospheric illuminations of 1908 had been due to the tail of a comet penetrating Earth's atmosphere.

Actually the cometary hypothesis, which would have better explained the nature of the Tunguska event, was not developed until the 1930s, though it could presumably account for the observed and reported "preparatory stage" – the atmospheric anomalies that preceded the event. Yet for the next decades, the enigma of these "Tunguska precursors" was almost forgotten. It was only in the early 1960s that Nikolay Vasilyev and other scientists brought the subject back to life when they carried out a detailed analysis of the anomalous atmospheric phenomena of the summer of 1908. In 1963, with the aid of the Rector of Tomsk Medical Institute, the Independent Tunguska Exploration Group (ITEG) sent out a questionnaire to most observatories that had existed in 1908 (to more than 150), asking colleagues both at home and abroad to report back on any natural phenomena that were recorded at their observatories in the summer of 1908. This was an ambitious project. Let's not forget that it was almost the climax of the Cold War and even postal contacts by Soviet citizens with foreigners were considered as suspicious by Party and State authorities. However, more than a 100 of the research bodies responded to the inquiries, and the agreement in the data received confirms its reliability. The ITEG researchers also read many Russian and foreign periodicals from the late 1900s for more first-hand information. They examined more than 700 Russian newspapers and journals, as well as the logbooks of ships that were at sea in the summer of 1908. The information collected was analyzed and the results published as the scholarly monograph *Noctilucent clouds and optical anomalies associated with the Tunguska meteorite fall.*[3] Even today, more than 40 years after its publication, that book is considered the most complete work on the subject.

So what conclusion did the scientists arrive at? As mentioned in Chapter 1, the strange atmospheric phenomena started as early as June 27, 1908. However, before June 30 they were observed only in certain places of western Europe, the European part of Russia, and western Siberia. The anomalies included unprecedented bright and prolonged twilights, an increase in the brightness of the night sky, and the formation of silvery clouds. In the early morning of July 1, these phenomena reached their peak, literally exploding in intensity and diversity. And throughout a territory of about 12 million km^2, there was no night separating June 30 and July 1 (see maps on Figures 2.1 and 2.2). How did these anomalies originate and why did they develop in this way? This remains a mystery, defying a final explanation, but later we will consider possible and probable solutions.

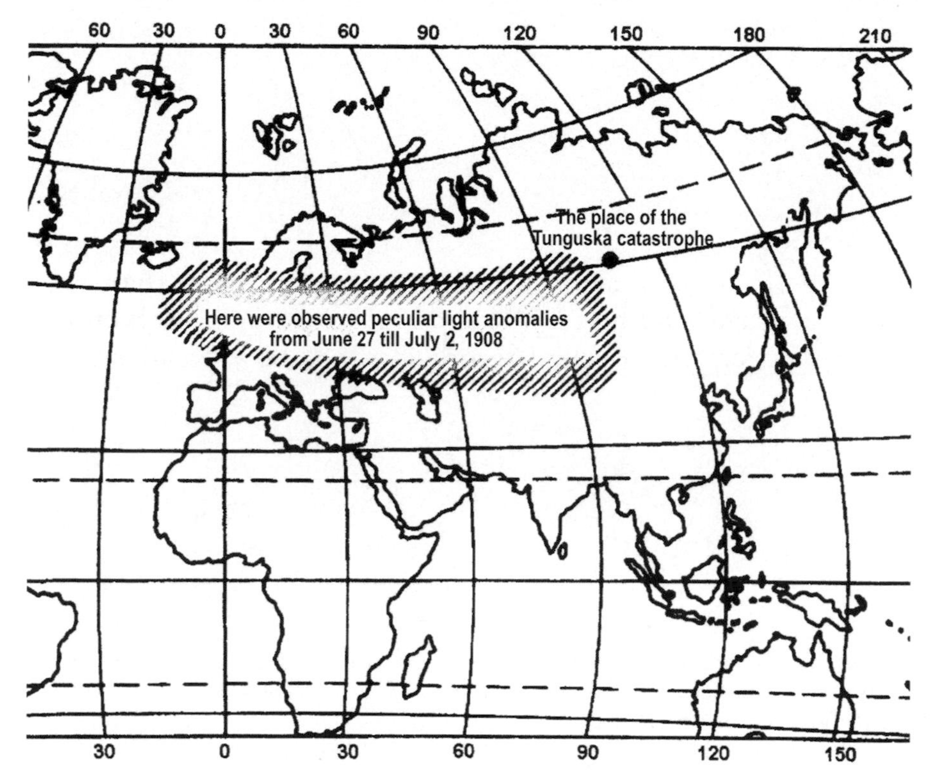

FIGURE 2.1. The region over which, from June 27 to July 2, 1908, peculiar light anomalies were observed in the atmosphere both before and after the Tunguska explosion (*Credit*: Vitaly Romeyko, Moscow, Russia).

In 1965, Nikolay Vasilyev and his colleagues at the ITEG analyzed information on the atmospheric phenomena that had been reported from 155 places of western, central, and eastern Europe, central Asia, and western Siberia. They found that until June 27, the twilight anomalies, even if reported, were few and far between. On June 29 they were seen in nine places, but on June 30 in more than 100 places. They then rapidly decreased (see diagram on Figure 2.3). Nothing like this had ever been seen before or since.

The journals and newspapers of those days reacted immediately to such amazing atmospheric phenomena. The St. Petersburg newspaper *Novoye Vremya* (*New Times*) of July 13 published an article by Sergey Glazenap, then professor of astronomy at St. Petersburg University, in which he described "light nights" that spread across regions of Russia. He said: "I have reports from several

FIGURE 2.2. Points from where especially intensive optical anomalies on the night of June 30–July 1, 1908, were reported (*Source*: Vasilyev, N. V., and Fast, N. P. Boundaries of the areas of optical anomalies of the summer of 1908. *Problems of Meteoritics*. Tomsk: University Publishing House, 1976, p. 126.).

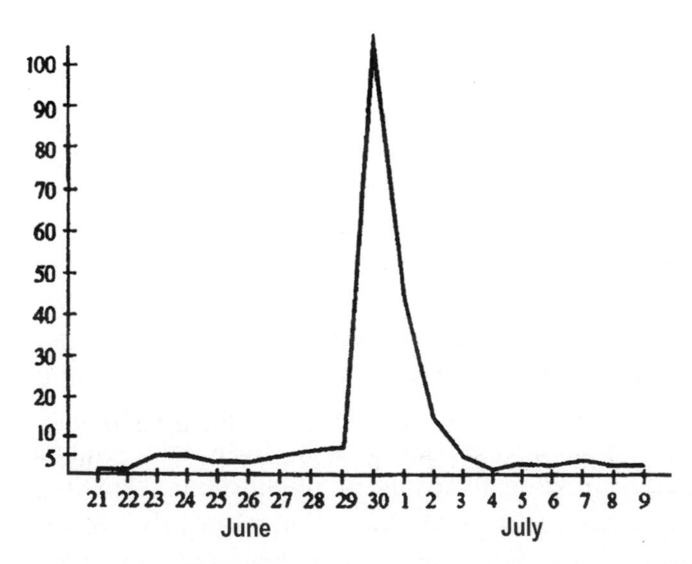

FIGURE 2.3. Diagram of the intensity of atmospheric optical anomalies in June and July of 1908 (*Source*: Vasilyev, N. V. *The Tunguska Meteorite: A Space Phenomenon of the Summer of 1908*. Moscow: Russkaya Panorama, 2004, p. 42.).

amateur astronomers about a phenomenon they believe to be northern lights. There was information in some newspapers about the Aurora Borealis occurring on June 30, and recently Mr. M. Taldykin from Lomzha sent me a detailed description of this night light, adding his opinion: 'northern lights, no doubt.' Yesterday, on July 10 here in Domkino, in the Luga district, after a rainy day the sky cleared up and the night was cloudless. I was then able to see the phenomenon myself, and I should state that it is quite different from a usual Aurora Borealis. It is rather a lucid twilight, similar to those observed in 1885 after the violent eruption of the Krakatoa volcano." [This is a misprint. The eruption actually happened in 1883.] Glazenap continues: "In Luga after sunset the northwestern part of the sky was intensely red. Far more than normal. By 10.30 pm the redness had disappeared, leaving behind a golden tinge so intense that when one looked at it the eyes could not bear its brilliance. This phenomenon lasted until midnight, when it began to weaken. It definitely resembled the red twilights we had in 1885, which were caused by the Krakatoa eruption, but the colors were much redder." So the conclusion here is that this was nothing like the Aurora Borealis.[4]

The Soviet astronomer Vasily Fesenkov was, in 1908, a student preparing in the evening of June 30 at Tashkent Observatory for his regular astronomical observations, but he waited in vain for night to fall. Nothing of this sort arrived.[5] In Heidelberg, the atmospheric phenomena over Germany were observed and described by Max Wolf, who reported that the sky after sundown became covered with unusual high-altitude cloudlets. They resembled cirri but were much higher than usual cirrus clouds. They looked rather like layers of smoke in the sky at sunset. The intensity of the nighttime luminosity was considerable. At midnight one could easily make out the hands and figures of a pocket watch. At 1.15 it was as light as daytime.[6]

The anomalies were reported from an area bounded by the Atlantic coast in the west, by the Yenisey River in the east, and by the Krasnoyarsk–Tashkent–Stavropol–Sevastopol–Bordeaux line in the south. Their northern boundary remained unknown. Amazingly, no atmospheric anomalies occurred in the area of Tunguska, which had its usual summer nights. There were observers in the area, but they did not see any. What this means remains unclear, even though some attempts to propose an explanation have been made by scientists in Russia and abroad. Anyway, the nearest point to the Tunguska event where the anomalies did appear was 600 km away.

In the town of Yeniseysk, Mrs. O. E. Olfinskaya, who lived there in 1908, later described her impressions: "Usually in this season (June 30) midnight in Yeniseysk is the darkest time of the day. But it was so light in the street that I was completely astonished. Other inhabitants of the town were also astonished. After an hour in the street I saw no sign of darkness."[7]

The intensity of the anomalies seemed to increase from East to West. In the very heart of Russia, in Kursk province, a local inhabitant, Mrs. Tomilina, had a similar experience to that of Mrs. Olfinskaya. "About 10 pm, after the evening twilight, it somehow became lighter instead of darker. The north-western part of the sky, and then the northern part of the horizon, brightened up as if just before sunrise, and soon everything was illuminated by a golden light. After a few minutes it got so light that one could read and discriminate things in their smallest detail. Even objects three to five kilometers away could be seen as distinctly as at dawn on a clear morning. Meanwhile an afterglow was flaring up in the north and north-east. A pale-azure sky on the horizon became golden and the clouds were tinged with pink. Then the sky was flooded with a crimson color. The unusual dawn woke birds. Poultry got upset and noisy. In the field quails were singing and flocks of awakened pewits took to the wing. About 11 pm the luminous phenomenon began to fade and had almost vanished by midnight, although the 'white night' lasted till morning."[8]

During these perplexing nights, in dozens of settlements across Europe and Russia, many photographs were taken of luminous clouds and buildings lit by this strange illumination. In 1991, the Russian astronomer Vitaly Bronshten estimated its brightness by examining these photographs. According to the photometric methods he used, the illumination was about a hundred times the normal brightness of the night sky.[9] In 1991, Vitaly Romeyko (a Moscow astronomer who took part in two dozen expeditions to Tunguska) used another method to estimate the brightness. He selected witness reports of the atmospheric anomalies and used 19 parameters that could be digitized, such as visibility of buildings, separate stars, the Milky Way, and printed notices shown in the photographs that could be read. The result is impressive: the level of the anomalous luminosity on the night of July 1 exceeded the nighttime norm by up to 800 times.[10] And, strange as it may seem, the highest levels were recorded far from Siberia.

The first analysis of the atmospheric anomalies of the summer of 1908 was actually carried out in 1908 by Alexander Schoenrock, Director of the Central Physical Observatory in St. Petersburg.[11] According to the data he analyzed, the night glow covered a quarter of the horizon. More often than not, it was an orange or reddish color, resembling the glow of a large fire, but sometimes it was evenly white or greenish. Schoenrock considered three explanations: first, the Aurora Borealis; second, a layer of thin high-altitude clouds illuminated by the Sun; and third, a penetration of dust into the upper strata of the atmosphere. None of these proved to be convincing enough. The first explanation seemed the least probable. The second looked somewhat more acceptable, but, as Schoenrock noted, the enormous territory on which the phenomenon was observed did not favor high-altitude clouds. Therefore, there remained the third possibility: increased dust in the atmosphere. But the fact that the imposing spectacle of light nights had completely stopped after two days did not support this explanation, either. At the time, of course, Schoenrock was not aware of the Tunguska event. So for him the atmospheric anomalies were just a strange phenomenon – especially as they ceased very quickly. In 1883, after the eruption of Krakatoa, unusually bright twilights had lasted several months, so how could dust from the Tunguska event disappear from the atmosphere so quickly? Obviously it could not have done so. And for current research on the subject, this seems to rule out the possibility that what happened at Tunguska was the fall of a usual meteorite, the impact of which, judging from the damage caused, would have put an enormous amount of dust into the atmosphere.

True, some decrease in the air's transparency in the summer of 1908 (through more dust being in the atmosphere) did in fact take place, but evidence of this was found only much later. In 1949, astronomer Vasily Fesenkov processed data for this period that the Mount Wilson Observatory in the United States had recorded. He concluded that a decrease in the transparency of the atmosphere not only took place, but was considerable, its magnitude and duration being unprecedented for the whole period between 1905 and 1911. It looked as if an enormous dusty cloud was moving over California in late July and early August of 1908.[12]

So the question is, did this cloud consist of the dispersed material from the Tunguska space body? Fesenkov believed it was probable, but the truth proved to be more complicated. In the 1980s, the Leningrad researcher Academician Kirill Kondratyev, an eminent Russian geophysicist and planetologist, along with Dr. Henrik Nikolsky and Edward Schultz, found that contemporary data showed that a decrease in the air's transparency because of dust had occurred in 1908, not only after but also before the Tunguska explosion. In that period scientists at the Astrophysical Laboratory of the Smithsonian Institute at Mount Wilson Observatory regularly measured levels of transparency of the atmosphere at various optical wavelengths. And for the first time – on June 4, 1908 – they detected an extensive dusty cloud that passed over Mount Wilson. Any decrease in transparency due to a higher level of dust almost a month before the explosion could hardly have had anything to do with the Tunguska space body. The dusty cloud detected in California continued to circulate around the globe with a period of 60 days while it gradually dispersed. But it appeared over Mount Wilson again on August 4 and on October 4.

Scientists calculated from the rate of the cloud's dissipation and the velocity of its motion through the atmosphere that it was formed from the impact of a large meteorite (mass no less than 100,000 tons) that had entered the atmosphere in the middle of May 1908 over the Pacific Ocean, not far from the Kuril Islands. It seems that due to the gentle slope of its trajectory, it did not hit the ocean but disintegrated in the atmosphere and completely burnt up, leaving behind a cloud of meteoritic dust. This meant there was no tidal effect that could have been observed. It was a normal meteor, one of many pieces of stone or iron that collide from time to time with Earth. It had nothing to do with the Tunguska space body.

But according to data on the optical density of the atmosphere measured by the Mount Wilson Observatory from July 14, 1908, there appeared over California yet another air mass that contained some strange substance. It was not dust.[13] The spectral signature of this substance, obtained in 1908 by Mount Wilson astronomers and processed in 1987 by Academician Kondratyev, does not correspond to dust but to an aerosol of ultramicroscopic particles suspended in the air. What is interesting here is that the date of its appearance in the United States is consistent with the time needed for such a cloud to travel from Central Siberia to California, so this substance could have been an actual product of the Tunguska explosion. And it could have been due to its aerosol composition that the optical atmospheric anomalies decreased so quickly after their culmination on July 1 (as distinct from similar cases of atmospheric dust from volcanic ash).

Alexander Schoenrock, at the Central Physical Observatory, who pondered in vain over possible explanations for the strange night glow, was both right and wrong at the same time: the dust did not disappear from the atmosphere because there was no Tunguska-related dust in the atmosphere. There was instead some other stuff whose nature still remains unclear, something that the reports from witnesses seem to confirm as the presence of a strange fluorescent substance in the atmosphere.

Alexander Polkanov, then a student but later a distinguished Soviet geologist, wrote in his diary in the summer of 1908: "A very unusual and rare phenomenon was observed in the night from June 30 to July 1 here, near the city of Kostroma. The sky is covered by a thick layer of clouds, and it is raining cats and dogs, but at the same time it is unusually light. It is already 11.30 pm but it is light, and it is still light at 1 am and is bright enough to read in the open. It can't be the Moon. The clouds are illuminated with a yellow-green light which sometime merges into pink. It is the first time I have seen such a phenomenon. As I watched I saw a layer of golden-pink clouds at a great altitude..."[14]

And that was not all. The nocturnal atmospheric anomalies of 1908 certainly looked spectacular; but apart from them there were the less-impressive daytime anomalies such as intense and prolonged solar halos, mother-of-pearl clouds, and a Bishop's ring. The so-called Bishop's ring, which is a diffuse brown or bluish halo around the Sun, occurs when there are large amounts of dust in the atmosphere. The first recorded observation of a Bishop's ring was made by the Reverend S. Bishop of Honolulu after the Krakatoa eruption. In Germany, W. Krebbs reported the presence of a Bishop's ring: "Starting from late June the light crown named after the

Reverend Bishop became a frequent associate of the Sun's disk during the first and last 15 min of its presence in the sky."[15] In another report, the same author provides a photograph of a Bishop's ring taken in Hamburg soon after June 30.[16]

Some meteorologists initially believed that all the atmospheric anomalies of June 27–July 2, 1908 were produced by a powerful volcanic eruption in a remote corner of our planet. However, investigations carried out both immediately after these phenomena and in the following decades by Russian and foreign specialists did demonstrate the fallacy of this explanation. Today the evidence indicates that these anomalies were directly related to the Tunguska event, which was not just a "local meteorite fall" and even something "more than regional."

The idea of a possible connection between the atmospheric anomalies of the summer of 1908 and the Siberian "meteorite" was suggested in 1922 to Leonid Kulik by Daniil Svyatsky, who was in the early 1920s the chief editor of the *Mirovedeniye* (*Cosmography*) journal.[17] But in 1908, neither Russian nor European scholars could find any such connection. It was even supposed that academics in the European part of Russia remained completely unaware of the event. However, in 2000, astronomer Vitaly Bronshten found that on September 25, 1908, the Russian newspaper *Sankt-Peterburgskiye Vedomosti* (*St.-Petersburg Records*) had told its readers about the fall of a huge meteorite in the Siberian taiga. And it was after reading this article that Permanent Secretary of the Imperial St. Petersburg Academy of Sciences, Sergey Oldenburg, became interested in the subject and had sent an official inquiry to the Governor of Yenisey Province, A. N. Girs – the nearest government official to the event. By that time, Girs had already received the report from the Yeniseysk District police officer I. K. Solonina about the bolide seen in the sky over Kezhma some 215 km from the place of the Tunguska explosion.

Solonina reported: "On the 30th day of June at 7 am in clear weather a bolide of enormous size flew at a great altitude over the village of Kezhma. It produced a number of loud sounds like gunshot reports and then disappeared..." But Mr. Girs for some reason feigned that he had no information on the Tunguska event. On October 10 he replied to Academician Oldenburg that he had ordered the Kansk District police officer S. G. Badurov to check the rumor about the bolide, that the official did investigate but could not confirm the rumor. Why the Governor behaved in this way remains unknown. Most probably he simply wished to avoid any complications. On October 21, 1908, the Physical and Mathematical Branch of the Academy of Sciences, after hearing an account of the alleged Siberian bolide, resolved to "make a note of the information," which meant that the question had been closed.

Well, Siberia is far from St. Petersburg (where the *St.-Petersburg Records* was published) and academicians did not then consider newspapers a reliable source of information, but the Siberian scientists of that period did not show their true worth, either. Soon after reports of the bolide's flight and the devastating explosion had appeared in local newspapers, geologist Professor Vladimir Obruchev, who then lived and worked in Tomsk, tried to check the newspaper reports but failed to find out whether the event they described had actually taken place. This may have been because he was 1,100 km from Vanavara, the settlement nearest to the Tunguska explosion.

However, it's difficult to be equally indulgent toward Arkady Voznesensky, the Director of the Irkutsk Magnetographic and Meteorological Observatory (see Figure 2.4). The manner in which he treated the information about the flight and explosion of the Tunguska space body collected by him in 1908 seems inexplicable. The observatory at Irkutsk had been established in 1884, and meteorological observations and magnetic measurements started there in 1886. Very soon the observatory became a leading geophysical center in Siberia. And in 1895 the noted geophysicist and climatologist Arkady Voznesensky became its director. Nobody would have called Voznesensky a conservative scientist. In 1907, he made two flights over Irkutsk in a balloon (a daring deed at the time), taking the first bird's eye photographs of the city and marking the beginning of regular aerial observations in that region. Equipment at the observatory was therefore always up to date. Voznesensky also created a special corresponding network of observers, aimed at collecting information about earthquakes, which were frequent in the region. This network included keepers of meteorological stations, postal employees, schoolteachers, and other representatives of the local intelligentsia. They could report earth tremors either on their own initiative or by filling out the forms that were sent from the observatory.

FIGURE 2.4. Dr. Arkady Voznesensky (1864–1936), Director of the Magnetographic and Meteorological Observatory at Irkutsk from 1895 to 1917, the first scientist who understood that a gigantic space body had entered the Earth's atmosphere and exploded over central Siberia (*Source*: Bronshten, V. A. *The Tunguska Meteorite: History of Investigations*. Moscow: A. D. Selyanov, 2000, p. 18.).

On the eventful day (June 30, 1908), two seismographs at the observatory recorded a weak tremor that was entered in "The List of Earthquakes Occurring in 1908." The tremor lasted from 0 h 19 min GMT to 1 h 46 min (see Figure 2.5). Two days before the Tunguska event, another tremor had been recorded that was more powerful and had a more normal signature of an earthquake. Arkady Voznesensky immediately sent out a questionnaire to his seismic network, asking his correspondents to provide details of these two earthquakes.

The director of the observatory, being totally unaware of the explosion at Tunguska, could have put nothing in the questionnaire to his seismic network that related to that event. He only asked questions about the characteristics of the two quakes. The first tremor (on June 28) was recorded by almost all of Voznesensky's correspondents. The second tremor – which was due to the

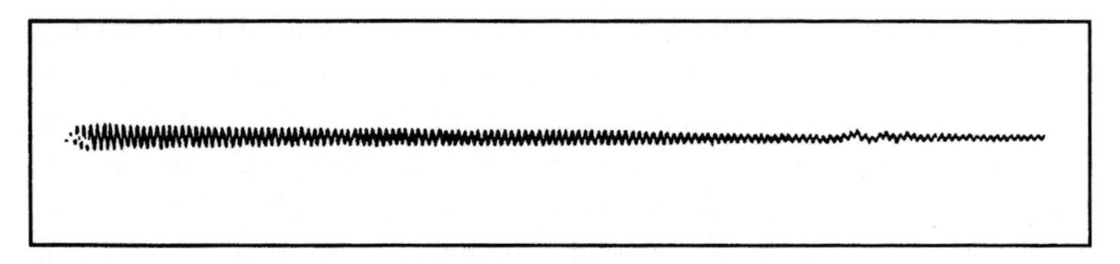

FIGURE 2.5. A seismogram of the Tunguska earthquake of June 30, 1908. These oscillations were produced by the explosion of the Tunguska space body and recorded by seismographs from the Irkutsk Magnetographic and Meteorological Observatory. Subsequently the Russian specialist in powerful explosions, Professor Ivan Pasechnik, used them to determine the exact moment of the Tunguska explosion (*Source*: Vasilyev, N. V. *The Tunguska Meteorite: A Space Phenomenon of the Summer of 1908.* Moscow: Russkaya Panorama, 2004, p. 86.).

Tunguska event – only by a few, although most respondents did hear sounds like thunder or the firing of large caliber pieces of ordnance on the morning of June 30. Sounds from the exploding Tunguska space body were heard in an area with a radius that exceeded 800 km from the epicenter, and some observers described a luminous body that could have been an enormous fiery meteor. The loudest sounds were reported by observers between the Lena and Yenisey rivers and Lake Baikal, although sounds were heard over an area of about 1 million km^2. The flying body was seen by 17% of those who replied to the questionnaire, all of them in the eastern part of the area. And 30% of the respondents reported the earth tremors.

Among the replies Arkady Voznesensky received, G. K. Kulesh at the Kirensk Meteorological Station wrote on July 6, 1908: "On June 30 to the northwest from Kirensk [a town some 500 km southeast from the site of the Tunguska explosion] local people observed an event that lasted from about 7.15 am till 8 am.[18] I myself could not see it, since having taken readings from my meteorological instruments I returned to the house and set to work. Although I did hear some thuds, I mistook them for gunshots from the nearby shooting-range. After work I looked at the barograph's band and noticed to my great surprise an additional line on the graph near the 7 am time marker, which indicated an abrupt and short jump in atmospheric pressure..."

Kulesh also reported on what local inhabitants had experienced. "At 7.15 am there appeared in the northwest a fiery pole like a spear about eight meters in diameter. The pole then vanished and one could hear five powerful abrupt and thunderous sounds. They followed each other quickly and distinctly. There then appeared at the same place a dense cloud. Within about 15 min one could hear similar thunderous sounds and still more 15 min later. A ferryman (a veteran soldier and clever man) counted 14 in all. Owing to his duties he was on the river shore, where he observed and heard the whole event from beginning till the end. Many people saw the fiery pole and even more heard the 'cracks of thunder.' Peasants from nearby villages came to the town and asked: What was that? Doesn't it betoken a war? They were told that an enormous meteorite had fallen. I should add that the 'cracks of thunder' came in three groups. As for the earth tremor, it was both felt and recorded by my barograph."

Mr. Kokoulin, an agronomist from the village of Nizhne-Ilimskoye, told Arkady Voznesensky in his letter of August 10: "On June 30, at about 7.15 am, workers who were building a bell-tower saw a fiery log flying from southeast to northwest. There were two sounds like gunshots followed by a very loud thunder and an earth tremor. The local people felt the earth trembling. One girl, a housemaid of a priest, fell down from a bench. People were afraid. Witnesses reported that clouds of black smoke rose like a pillar where the space body fell – or rather where it went below the horizon. The Tungus people who wandered behind the settlement of Nizhne-Karelinskoye (to the west-northwest from Kirensk) say that there were terrible crashes of thunder..."

A. A. Goloshchekin, living in the village of Kamenskoye (about 600 km west-southwest from the explosion site), reported in his letter of June 30: "At 7 am in this village there were three succeeding underground thunderclaps from a northwestern direction. At the same time people felt an earth tremor. From questioning local inhabitants I learnt that several minutes earlier they saw a flying oblong body that narrowed towards one end. It seemed as if the body had broken away from the sun, for its head was as bright as the Sun while the remaining part was a misty color. The body, having covered some distance, fell in the northeast."

It's unfortunate that the questionnaire sent out by Voznesensky at the Irkutsk Observatory was aimed at collecting information only about seismic phenomena, and did not ask questions about the direction and angular heights of the bolide's flight, or the flight itself. Some respondents did report the bolide's flight, but others who saw it may have refrained from mentioning it, either because they were not asked about this directly or from fear of ridicule. But anyway, the data Voznesensky collected, being obtained very soon after the event, are definitely the most important initial source of information about the Tunguska space body. He processed the data and determined, using readings from the seismometers, that the probable coordinates of the body's fall were 60°16′N, 103°06′E, and the probable time of the fall as 0 h 17 min 11 s GMT.

So, Voznesensky in 1908 had achieved an enviable precision in his calculations that were based mainly on the reports of witnesses. He also calculated that the trajectory of the Tunguska space body was from south-southwest to north-northeast. What seems astounding is that Voznesensky at once understood that the Tunguska space body did in fact explode in the air, even if he called this process the "rupture of the meteorite" and overestimated the altitude of the explosion by a factor of three. (In the 1970s, the altitude of the Tunguska explosion was determined fairly accurately by several methods at somewhere between 6 and 8 km.) Voznesensky thought the meteorite had broken into pieces at the height of 20 km and that fragments then fell to the Earth's surface to produce the tremors that were reported. This informed guess was going to be rather important. But the main discovery that he made was the association between two seemingly unrelated facts: the earthquake tremors and the arrival of the space body.

An account of what was thought to be a weak earthquake in central Siberia on June 30, 1908, was presented to the Seismic Committee of the Imperial Academy of Sciences. However, Arkady Voznesensky did not dare include any information about the flight of a huge bolide, or his calculated coordinates of the epicenter of its explosion. Igor Astapovich, a Ukrainian astronomer, once said that Voznesensky feared his report would have looked "fantastic."[19] Only in 1925 did he decide to publish the data.[20] But by then it was too late for him to become *the* pioneer of Tunguska studies. This title already belonged to Leonid Kulik.

It was Kulik who ventured to believe in the testimonies of witnesses and newspaper articles, while at the same time being unaware of the instrumental detection of the Tunguska event at the Irkutsk Observatory. However, the "meteorite hypothesis" for the space body was not authored by Kulik. And whether this hypothesis is correct still remains doubtful. An iron meteorite had definitely been rejected as a possibility, but some specialists believed that a *stony* meteorite could explode in the air and produce all the Tunguska effects. Yet others strongly disagreed, proving mathematically that it was impossible and pointing out that if a stony meteorite had exploded the whole place would have been strewn with its remains. And after many expeditions to the site, nothing like this sort of evidence had been discovered in the Tunguska taiga. In any case, the very word "meteorite" was first used by Siberian newspaper reporters who were in no way noted for their scientific accuracy, though they didn't fear to tell the public what they saw and heard, or to use such a term as "a huge meteorite" – which the distinguished scientist Arkady Voznesensky decided against doing.

Perhaps in the data Voznesensky collected there was "something more," something that did not fit the accepted view of meteorites, and that "something" he decided to keep to himself. We're guessing of course, but the unnatural behavior of this Russian geophysicist provides good reason to mention such a possibility. To instrumentally record an earthquake produced by a meteorite fall (for the first time in history!) and to gather data from professional observers to determine the probable coordinates of the meteorite's fall are remarkable. Using the data he possessed, Voznesensky could have written an important scientific paper that would have been accepted for publication by any scholarly periodical of the time. After all, by 1908 the study of meteorites had become a completely legitimate discipline within science. Meteorites were an accepted part of the Solar System – and they often hit Earth. If Voznesensky's paper of 1925 had been published in 1908, there would have been no reason to blame him for an unscientific approach to the event. But he postponed writing that paper for 17 years.

A normal scientist – and Arkady Voznesensky was quite normal – could not have acted in such a manner without a real reason. So did Voznesensky – not being a specialist on meteorites – think it best to refrain from expressing his opinion? Hardly so. In his time, scientists were not as narrowly specialized as they later became. And meteoritics itself was still in its infancy. It was astronomers, geologists, chemists, and geophysicists who were participating in this new branch of science. In 1925 (two years before Kulik reached the site of the Tunguska meteorite fall), nobody said Voznesensky's paper was unscientific. Colleagues actually expressed their regret that such a paper had been published so late.[21]

Of course, we may have underestimated the power of scientific conservatism. A meteorite of decent dimensions would have been a respectable subject for a scholarly paper, but a *gigantic* meteorite...? This had the smell of a sensational newspaper story. Besides, as we know, several minutes after the explosion, a local geomagnetic storm began that the instruments at Voznesensky's observatory recorded. The director could hardly have missed the strange coincidence of this magnetic storm. And its significance must have puzzled this noted geophysicist. So, it could have been this strange geomagnetic disturbance from the explosion that made him keep back the recorded data from the St. Petersburg academic authorities and from the scientific community as a whole. This, of course, is only one explanation for why Voznesensky might have kept things to himself. Even on its own, the very first post-meteoritic earthquake was quite a discovery. But if one adds that the first and the last post-meteoritic geomagnetic storm was also recorded, one can begin to see why this may have been too much for the science community of the time. And let's not forget the widespread nighttime illuminations, the nature of which remained far from clear and might have had something to do with the Tunguska event. Silence is sometimes more expressive than words, and the fact that Voznesensky's paper of 1925 completely ignores both the optical atmospheric anomalies and the geomagnetic storm of June 30, 1908, is intriguing. But he has taken this mystery with him to the grave.

Anyway, judging from his paper of 1925, Arkady Voznesensky had no doubts that the Tunguska space body had been a meteorite. The only thing for him was the choice between a stone and an iron meteorite. He wrote: "There is a good probability that a future investigator of the site where the meteorite has fallen will find there something similar to the Arizona meteor crater." His prediction was wrong, but at least his mistake was excusable – as distinct from his dead silence in 1908. For if we are trying to find the *main* reason for the oblivion into which the subject of the Tunguska meteorite had almost fallen, it was certainly because of the extreme scientific caution of the director of the Irkutsk Magnetographic and Meteorological Observatory. All other factors (such as the remoteness of the area of the meteorite fall or even the prevarication of Mr. Girs, the Governor of Yenisey Province) were far less significant. Was his silence due to the very strange and inexplicable geomagnetic effect that accompanied the explosion? Nobody at present can say, but if it was so, this provides another paradox in the Tunguska story. For, as we will see, the geomagnetic effect is perhaps the most specific and unusual aspect of the whole subject.

Notes and References

1. See Svyatsky, D. O. Unusual twilight on the night of June 18 (old style) in Tambov. – *Astronomicheskoye Obozreniye*, 1908, No. 6; Svyatsky, D. O. Illumination of twilights. – *Priroda i Lyudi*, 1908, No. 37.
2. See Vasilyev, N. V. *The Tunguska Meteorite: A Space Phenomenon of the Summer of 1908.* Moscow: Russkaya Panorama, 2004, pp. 38–50 (in Russian); Schoenrock, A. M. The twilight of June 17 (30), 1908. – *Meteorologichesky Bulletin Nikolayevskoy Glavnoy Geofizicheskoy Observatorii*, 1908, No. 6 (in Russian); Suering, R. Die ungewohnlichen Dammerungserscheinungen im Juni und Juli 1908. – *Berichte der Preußischen Meteorologischen Instituten,* 1908; de Roy, F. Les illuminations crépusculaires des 30 juin en 1 juillet 1908. – *Gazette astronomique*, 1908, No. 8.
3. See Vasilyev N. V., et al. *Noctilucent clouds and optical anomalies associated with the Tunguska meteorite fall.* Moscow: Nauka, 1965 (in Russian).
4. See Glazenap, S. Light twilights. – *Novoye Vremya*, 1908, No. 11601 (in Russian).
5. See Krinov, E. L. *The Tunguska Meteorite*. Moscow: Academy of Sciences of the USSR, 1949, p. 84 (in Russian).
6. See Wolf, M. Uber die Lichterscheinunsen am Nachthimmel aus dem Anfans das Juli. – *Astronomische Nachrichten*, 1908, Bd. 178.
7. See Krinov, E. L. op cit., p. 82.
8. See Tomilina. Description of a light phenomenon that occurred on June 17, this year, in the Tim district of Kursk province, as well as in the Manturov settlement and other places of the same province. – *Priroda i Lyudi*, 1908, No. 37 (in Russian).

9. See Bronshten, V. A. The nature of the anomalous sky luminance related to the Tunguska phenomenon. – *Astronomichesky Vestnik*, 1991, Vol. 25, No. 4 (in Russian).
10. See Romeyko, V. A. On the nature of the optical anomalies of the summer of 1908. – *Astronomichesky Vestnik*, 1991, Vol. 25, No. 4 (in Russian).
11. See Schoenrock, A. M. The twilight of June 17 (30), 1908. – *Meteorologichesky Bulletin Nikolayevskoy Glavnoy Geofizicheskoy Observatorii*, 1908, No. 6 (in Russian).
12. See Fesenkov, V. G. Turbidity of the atmosphere produced by the fall of the Tunguska meteorite of June 30, 1908. – *Meteoritika*, Vol. 6, 1949 (in Russian).
13. Kondratyev, K. Y., Nikolsky, H. A., and Schultz, E. O. The Tunguska space body – the core of a comet. – *Current Problems of Meteoritics in Siberia*. Novosibirsk: Nauka, 1988, p. 126 (in Russian).
14. Polkanov, A. A. About phenomena accompanying the fall of the Tunguska meteorite. – *Meteoritika*, Vol. 3, 1946 (in Russian).
15. Krebbs, W. Die Liechterscheinungen am Nacnthimmel das 30. Juni 1908. – *Das Weltall*, 1908, 9 Jahrgang, Heft 1, 1 Oktober.
16. Krebbs, W. Photographien der Nachammerung des 30 Juni 1908 und einer Bishopschen Aureole. – *Meteorologische Zeitshrift*, 1908, Bd. 27.
17. See Kulik, L. A. About a possible connection between meteorites and comets. – *Mirovedeniye*, 1926, Vol. 15, No. 2, p. 174 (in Russian).
18. This included all the phenomena: the bolide's flight, sounds, tremors and so on.
19. See Astapovich, I. S. About the possible trajectory and orbit of the Tunguska comet. – *Physics of Comets and Meteors*, Kiev: Naukova Dumka, 1965, p. 107 (in Russian).
20. See Voznesensky, A. V. The fall of a meteorite on June 30, 1908, in the upper reaches of the river Khatanga. – *Mirovedeniye*, 1925, Vol. 14, No. 1 (in Russian).
21. See, for example, Astapovich, I. S. The Great Tunguska meteorite. Part 1 "History of investigations". – *Priroda*, 1951, No. 2 (in Russian).

3. A Shocking Discovery

Soon after its spectacular flight and devastating explosion over the Siberian wasteland, interest in the Tunguska space body practically evaporated. Turbulent times were approaching, and cosmic stones began to look less important. With war breaking out in Europe in 1914 and all that happened afterward, there was much to keep the science community from exploring the Tunguska catastrophe. That war proved to be a turning point that determined the catastrophic nature of the twentieth century. If there had been no war there would have been no October Revolution of 1917 in Russia, and history would have followed a very different path. If we believe in the "many-worlds" interpretation of quantum mechanics, we can suppose that "somewhere" a better world history has materialized. But not here, alas.

One participant in that war was the mobilized student of the Mineralogical Faculty of St. Petersburg University, Leonid Kulik, who became the future pioneer of Tunguska studies (see Figure 3.1). By that time Kulik was already 30, with mineralogy the passion of his life. He was born on September 1, 1883, in the Russian town Derpt (now the Estonian town Tartu). His family belonged to the gentry, although they were not rich, and after the early death of his father the family moved to Troitsk in the Urals. Here, in 1903, Leonid Kulik gained a gold medal at the Troitsk Classical Grammar School and entered the St. Petersburg Imperial Forest Institute, where he was influenced by the craze for "leftish ideas." A year later, in 1904, he was expelled from the institute for taking part in student disturbances and was called up for military service. But a military career was not for him, and the stormy year of 1905 found Kulik participating in an armed revolt in Kazan. The revolt was suppressed, and Kulik soon ended his military training and returned to Troitsk in the Ural Mountains, which is probably the most suitable place in the world for a lover of stones.

V. Rubtsov, *The Tunguska Mystery*, Astronomers' Universe,
DOI 10.1007/978-0-387-76574-7_3, © Springer Science+Business Media, LLC 2009

FIGURE 3.1. Dr. Leonid Kulik (1883–1942), the pioneer of Tunguska studies (*Source*: Krinov, E. L. *The Tunguska Meteorite*. Moscow: Academy of Sciences of the USSR, 1949, p. 4.).

While working at the Mining department, Kulik studied mineralogy as well as botany and zoology. At heart, Kulik was a naturalist and an empiricist, a devout successor of those who studied lightning, meteorites, and volcanoes and created herbaria – quite unlike modern theoreticians and experimenters. Another of his passions seems to have been underground work for the Revolution. In 1911, police arrested him, and he spent three weeks in the Troitsk citadel, which was used as a prison. He was hardly an "innocent victim" of the Tsarist regime, but his guilt was not established and he was released, although he remained under police surveillance. The country still had laws, not all of which were draconian. For some time Kulik worked as a forest warden, but the path of his life changed abruptly when he met a member of the Imperial St. Petersburg Academy of Sciences, Dr. Vladimir Ivanovich Vernadsky, a famous geochemist and authority on radioactivity (see Figure 3.2). The

FIGURE 3.2. Academician Vladimir Vernadsky (1863–1945), an eminent geochemist and inspirer of Tunguska investigations in the 1920s and 1930s (*Source*: Zhuravlev, V. K., Rodionov, B. U. (Eds.) *Centenary of the Tunguska Problem: New Approaches.* Moscow: Binom, 2008, p. 418.).

subject of radioactivity in the first decade of the twentieth century was a hot topic in science – too hot for some researchers who died from studying it – and Vernadsky, being attracted to new lines of inquiry, had been trying to broaden scientific investigations in the field of radioactivity.

The Academy of Sciences listened to Vernadsky and decided to allocate funds for his expeditions to look for radioactive minerals. So, in the spring of 1911, with some colleagues, he visited the Caucasus and the Ural Mountains. The expedition needed a specialist in geodesy, someone who could determine exact geographical positions for the expedition, and the chief of the Mining Department recommended Leonid Kulik. Thus Kulik met Vernadsky, and their long association and joint research work commenced. Later this proved to be of considerable importance for the problem of the Tunguska meteorite. Without Academician Vernadsky's support, Kulik would hardly have succeeded in organizing his expeditions to Tunguska.

From August 20, 1912, Kulik was on the staff of the Academy of Sciences. This helped the Ministry of Internal Affairs to exonerate him from his former political charges and allowed him to live in both capitals of the then Russian Empire – in St. Petersburg and Moscow. Immediately he moved to St. Petersburg, where he cataloged minerals at the Peter the Great Geologic and Mineralogical Museum until the war in 1914 interrupted his studies. He enlisted in the engineer battalion of a cavalry brigade, the Dragoon Regiment of Finland (then part of Russia) that took part in some bloody battles in eastern Prussia. Kulik was decorated for bravery and later made a lieutenant.

In July 1917, the Provisional Government of Russia began to realize that the country needed the specialists who were perishing in the trenches of the Great War, and Kulik was recalled from Field Forces to St. Petersburg (which had been by that time renamed Petrograd). He then enlisted in the Central Scientific and Technological Laboratory of the War Ministry. Although the coup of October 1917 had been welcomed by Kulik as a "long-awaited victory," that victory turned into years of almost biblical calamities. The strife of Civil War brought Kulik into various regions of the country. He evacuated his family from starving Petrograd, looked for ocher in the Ural Mountains, taught mineralogy in Tomsk University, served initially in the White Army and then in the Red Army (in both cases for a short time), and again taught mineralogy in Tomsk.

Kulik's first encounter with the arrival of a new meteorite occurred at the beginning of the Civil War in April 1918, when a "sky stone" fell near the town of Kashin. The Academy of Sciences commissioned him to discover the circumstances of this event and to bring the meteorite back to Petrograd. Alas, the stone itself had already been sent to Moscow by the local authorities, although when Kulik arrived he obtained small fragments for the Academy. But his work was temporarily terminated for the next three years while his country, with a revolution and the Civil War, was in no mood for meteorites.

Eventually the Civil War came to an end, and in March 1921 Leonid Kulik returned to Petrograd to the post of Secretary to the Meteoritic Department at the Mineralogical Museum, which was headed by no less a person than his friend and mentor Vladimir Vernadsky. And a few days after his return an event occurred that changed the course of his life. "How distinctly I remember that moment," wrote Kulik seven years later. "It was March 1921 and Daniil Svyatsky, the Editor of the *Mirovedeniye* journal, approached me with an old page from a wall calendar dated July 2, 1910. 'Look at the back of this page,' he said. 'It is rumored that a giant meteorite fell in Siberia in 1908 near the Filimonovo railway station. And you know there's no smoke without a fire'."[1] Both these men proved to be very perceptive, Svyatsky because he recognized significant data in an old calendar and Kulik because he realized that he might follow up on its contents and make important discoveries.

As a matter of fact the calendar had a reprint of the most fictitious newspaper report on the Tunguska phenomenon, which Leonid Kulik called the "Filimonovo meteorite" after the railway station of Filimonovo. The journalists on the *Siberian Life* of July 12, 1908, had grabbed the public's attention with the title "A Visitor from Heavenly Space." The article told of a huge hot meteorite that had fallen near the station at Filimonovo and that eyewitnesses and scientists had examined it. The only doubt expressed by Kulik was that "its size might have been exaggerated by the author of the article." But he thought the story itself had been based at least partly on facts. Kulik went on: "The author gave the very natural circumstances of the meteorite fall as well as its exact date and place. Therefore, it can hardly be considered idle fantasy of a smart journalist to arouse our mystification." In fact the article was almost nothing but "idle fantasy." In retrospect, one can congratulate the reporter Alexander Adrianov, since his ability to compose fantastic stories helped to stimulate interest in the Tunguska event and encourage future expeditions and research.

The Russian Society of Amateurs of Cosmography, and the editor of its journal the *Mirovedeniye*, became most important in collecting and promoting information on the Tunguska event when the subject was almost forgotten. But its editor, Daniil Osipovich Svyatsky (see Figure 3.3), suffered cruelly at the hands of the Soviet authorities and the Society was disbanded in 1930. Its many

FIGURE 3.3. Daniil Svyatsky (1881–1940), a Russian historian of astronomy, the chief editor of the *Mirovedeniye* ("Cosmography") journal, who enthusiastically supported the search for the Tunguska meteorite in the 1920s (*Source*: Bronshten, V. A. *The Tunguska Meteorite: History of Investigations*. Moscow: A. D. Selyanov, 2000, p. 80.).

members were sent to gulags and Daniil Svyatsky was arrested in the spring of 1930 and kept in prison for many months. He was accused of being a secret monarchist because he had proposed naming a nova star that became visible in 1670 in the constellation of Vulpecula after the Russian emperor Peter the Great. For this, Svyatsky was condemned and sent with other State convicts to build a canal from the White Sea to the Baltic. In less than two years some 100,000 of these political prisoners had perished, though Svyatsky survived to be released in 1932. He then lived in Leningrad but was exiled to Alma Ata in 1935, when the authorities started a witch hunt for purported conspirators against Sergey Kirov, a noted member of the *Politburo* who was murdered in December 1934. This outstanding Russian historian of astronomy never returned from exile. He died in January 1940 when only 58. Although late in the 1920s and afterward, the leading part in Tunguska studies was played by the Academy of Sciences, it was the *Mirovedeniye* journal in the 1920s that held most information about the phenomenon and also argued for an expedition to investigate the place where the meteorite had fallen.[2]

Fortunately in that period several influential members of the Academy of Sciences, including Academician Vladimir Vernadsky, conceived a plan to organize the first large expedition through Russia to collect meteorites. (And it was through Russia because before December 1922 there was no Soviet Union, although the various states were moving toward forming such a union.) By that time, the academic archives contained many reports about meteorite falls in various parts of the country. So, on April 20, 1921, a meeting of the Physical and Mathematical Branch of the Academy of Sciences took place at which Vernadsky read a report prepared by Leonid Kulik entitled "New data about meteorite falls in Russia." The state of affairs in Russia at the time hardly favored the planned expedition. According to Kulik, the Academy of Sciences had no funds for it, while the "scientists themselves were emaciated and ragged."[3] Nevertheless, thanks to the support of the People's Commissar of Public Education, Anatoly Lunacharsky, the government allocated funds from the state budget.

The expedition led by Leonid Kulik numbered some 20 people, and with a private railcar they left Moscow on September 5, 1921. Searching for the "Filimonovo meteorite," as Kulik had labeled it, was not the only purpose of this trip, but the search did start in central Siberia in the town of Kansk, where the scientists distributed some 2,500 questionnaires to local inhabitants, hoping to collect information about what happened on June 30, 1908. While visiting the station at Filimonovo, Kulik concluded that no meteorite had ever fallen there, though the information gathered by the expedition proved that the rumor about the "giant meteorite" was not groundless.

As Kulik reported to the Academy of Sciences: "At about 5–8 am, June 30, 1908, an impressive meteorite flew over Yenisey Province from the south to the north and fell near the Ogniya River... The fall was accompanied by a brilliant light, a small dark cloud, and some very loud claps of thunder. The catastrophic impact of the leading air wave must be emphasized because according to reports from the Tungus it not only broke and felled many trees but also dammed the Ogniya River, having brought down the riverside cliffs."[4]

The expedition also investigated other meteorite falls in Siberia and the European part of Russia before returning to Petrograd on October 19, 1922. It lasted more than a year and covered some 20,000 km, gathering for the Mineralogical Museum specimens from ten meteorite falls.

Nevertheless, Kulik and his team did not reach the actual area of the Tunguska event, being aware that it would be impossible to get there without more extensive preparations. So, the material collected by the expedition provided only indirect evidence and evoked a skeptical reaction from many academics. Eyewitness reports (especially from native inhabitants of Siberia) about the flight and explosion of a "brilliant body" appeared to them scientifically worthless and did not justify more funds for an expedition to the place where this body fell. For several years, Leonid Kulik regularly submitted applications for another expedition, and the Academy refused his requests no less regularly. The absence of Vladimir Vernadsky, who was at this time lecturing at the Sorbonne and conducting experiments in French laboratories, also seems to have been a negative factor.

But in 1925 the situation began to improve. The *Mirovedeniye* journal published an article called "About the place of the 1908 Great Khatanga meteorite fall."[5] The article was by geologist Sergey Obruchev,[6] the son of the geologist and investigator of Asia, Vladimir Obruchev (mentioned earlier), who also wrote some of the most popular science fiction novels in the first half of the twentieth century. (English translations of his "Plutonia" and "Sannikov Land" are still available today from bookstores and the Internet.) When living in Tomsk in 1908, Vladimir Obruchev had tried to verify the newspaper reports about the Tunguska meteorite immediately after the event but had failed.

But back to his son Sergey Obruchev, who in 1924 was sent by the Geological Committee to examine geological features of the region by the Podkamennaya Tunguska River. Here he happened to discover that the fallen forest area of the Tunguska event was not far away. He wanted to visit the site but failed to persuade any Tungus guides to accompany him. According to Obruchev they "flatly denied that a meteorite had fallen." As Obruchev said in

1925: "The lack of time and means did not allow me to make a survey of such a large space covered by dense forest. Therefore, I had to restrict my investigation to collecting new eyewitness reports." In fact, these "new eyewitness reports" contained no new information about the Tunguska event, but they confirmed what was already known from the newspaper publications of 1908 and the work of the Meteoritic Expedition of 1921. Nonetheless, Obruchev's report prompted Arkady Voznesensky, the leading figure in the subject at the time, to publish a paper about the instrumental data obtained at his Irkutsk Observatory way back in 1908, which had confirmed that a large space body had fallen in Central Siberia.[7] Consequently, the contributions from Obruchev and Voznesensky greatly strengthened Leonid Kulik's position in scientific society, even though it did not influence the Academy of Sciences to finance a new Siberian expedition.

Soon, however, a new personage in the form of Innokenty Mikhaylovich Suslov (see Figure 3.4) entered the Tunguska

FIGURE 3.4. Innokenty Suslov (1893–1968), an anthropologist, the Chairman of the Krasnoyarsk Committee for Assistance to Northern Peoples, and one of the pioneers of Tunguska studies (*Source*: *The Tunguska Phenomenon: 100 years of an unsolved mystery*. Krasnoyarsk: Platina, 2007, p. 16.).

community. He was an anthropologist and a representative of Soviet power in Siberia. He first heard about the Tunguska catastrophe in the autumn of 1908, when a student at the local gymnasium. And helped by his teacher, the young Innokenty tried "to determine the location of the meteorite fall (or explosion) and to find out how it would be possible to get there."[8] The extraordinary event remained in Suslov's memory, and in March 1926 he questioned some Tungus people (now known as Evenks) who, at the moment of the Tunguska explosion, were near its epicenter. This new information, which contained important details, had been missed by newspaper reporters and by Kulik's Meteoritic Expedition of 1921–1922. In particular, Suslov talked with brothers Chuchancha and Chekaren (whom we met earlier), who described to him the sequence of several flashes and explosions over the Tunguska taiga.

Suslov's article "The search for the great meteorite of 1908," which was based on his talks with numerous Evenks, was published in the *Mirovedeniye* journal. It again confirmed the flight of a space body over central Siberia in 1908 as well as the probable location of the fall. However, another expedition to this region would have probably been postponed again had not Vernadsky returned from abroad and insisted on organizing one. So in February 1927 Kulik and his assistant Oswald Guelich left Leningrad for Siberia. In the middle of March they reached the Angara River and traveled downstream to the old Russian village of Kezhma, then occupied by *starovers* (old believers who escaped religious persecution after the church reform in 1655 and 1656). Here they obtained more information about their route and left for Vanavara, the village that was 70 km from the Tunguska event and the closest to it. They arrived at Vanavara on March 25.

On arrival, Kulik hired a guide – not without difficulties because the Evenks didn't want to visit places declared forbidden by their shamans. However, an Evenk named Luchetkan did agree to take them on horseback to the site of the meteorite fall, but the snow was still too deep for horses and they were forced to return to Vanavara. This made Kulik and his companion realize why the Evenks preferred deer to horses for their transport. A herdsman named Okhchen, the owner of a dozen deer, then agreed to help the expedition, providing his services were paid for, and on April 8 the travelers started out again. Apart from Kulik and Guelich, there was the herdsman Okhchen who took his younger wife, daughter, nephew – and even his baby. Five days later they entered the area of fallen wood (see Figure 3.5). Kulik described the scene: "All large trees on the mountains were leveled in dense rows, whereas in the valleys one could see both roots and trunks of age-old giants of the taiga broken like reeds. The tops of the fallen trees were directed to us. We were going north towards the super-hurricane that had raged here almost 20 years ago."[9]

FIGURE 3.5. The forest completely leveled by the shock wave of the Tunguska explosion. The photograph was taken in 1929, by Evgeny Krinov (*Source*: Krinov, E. L. *Foundations of Meteoritics*. Moscow: Gostekhizdat, 1955, p. 99.).

On April 15, Kulik climbed the Shakrama Mountain and for the first time saw the unbelievable "Land of Dead Forest." "I am still unable to sort out the chaos of the impressions that I took from that excursion," Kulik wrote in his diary, "and I even cannot imagine the whole colossal scale of this extraordinary meteorite fall. Here is a very hilly, almost mountainous locality, extending for tens of kilometers behind the northern horizon. Distant mountains along the Khushmo River are covered by a blanket of snow half a meter deep. And from our observation point one can see no sign of living forest:

everything has been leveled and scorched, but around this dead area a young (not older than 20 years) growth appears, striving towards the sun and life… It is so terrible to see the giants with a diameter ten to twenty *vershoks* [up to one meter] broken in two like a thin reed with their tops thrown aside for many meters to the south."

The aim of the expedition seemed to have been attained. But it only seemed so. By that time herdsman Okhchen happened to remember the shamans' ban on visiting this area and flatly refused to go further, and on April 19 the travelers had to begin their return. The Evenks were so eager to leave the forbidden area that the deer caravan got back to Vanavara in just two days.

Being disillusioned with the Evenks, Kulik decided to make arrangements with some Russian settlers living by the Angara River. Two hunters helped the scientists to build an intermediate camp on the Chamba River about 75 km from Vanavara, and the expedition members knocked up two rafts for nine people – and a horse. The horse sometimes pulled the rafts and sometimes traveled on them. It was spring, and the Chamba River was seething, but having reached the mouth of the Khushmo River they moved upstream, their one horse towing the two rafts.

On May 30, the expedition arrived at the mouth of the Churgim Creek, which provided too little water for a boat or a raft. The expedition set up "Camp No. 13" nearby, from which they began their examination of the surrounding area. They soon found to the north of the camp a vast hollow surrounded by mountains, which Kulik named the Great Hollow.[10] He then surveyed the directions of the fallen trees within the Great Hollow and discovered to his surprise that the whole forest had been put down in a radial manner.

"On a mountain pass," wrote Kulik, "I made my second camp and began to circle around the Great Hollow, passing by the mountains. First I went to the west and covered tens of kilometers by lonely mountain ridges, but always the fallen trees were oriented to the west! Then I circled the hollow to the south and the fallen trees, as if enchanted, turned to the south as well. I returned to my camp and went further by mountain slopes, now to the east, and the leveled trees started to shift their tops in the same direction. Finally, straining every muscle, I moved to the south once again, almost reaching the Khushmo River, and the lying bristle of the fallen wood turned to the south as well... There could be no doubts: I had circled the center of the fall."[11]

Finding the radial pattern of the fallen forest around the epicenter of the Tunguska explosion was an opening shot in the whole of Tunguska studies. Of course, at that time, Kulik did not use the term "epicenter." He believed that the meteorite – as any normal stone or iron meteorite would have done – did fall into the "Great Hollow." As he wrote, "It is with a fiery jet of burning-hot gases and cold bodies that the meteorite struck the hollow with its hills, tundra, and marsh..." He was completely sure that this pictured the event of 1908. But even though this proved to be wrong, it was Leonid Kulik who discovered the only area of radially leveled forest existing on our planet.

Unfortunately or fortunately, depending on one's viewpoint, there were in the northeastern part of the hollow several dozen flat craters similar to lunar craters. Naturally, Kulik, who was looking for evidence of a giant meteorite, decided that they had been formed by the fallen pieces of the space body. Later on, when the non-meteoritic nature of these craters was convincingly proved, some armchair researchers hurled plenty of unfair accusations at the pioneer of Tunguska studies. But what else should he have thought, having got to the place of the catastrophe and seen these craters? Yes, Kulik did make a mistake – but it was a "happy mistake." If he had understood at once that these craters were simple thermokarst holes, formed in this region when ice-rich permafrost melted, he could have decided that the Tunguska meteorite had fallen at another place and that the leveled forest was due, say, to an "untypical hurricane." In this case, Kulik would have started a long fruitless search for this "other place" – since he was a specialist in meteoritics looking for a meteorite and not for traces of hurricanes. The real "mega-trace" of the Tunguska explosion was the taiga itself, with its radially leveled trees over an area of some 2,100 km^2, which suggested a high-altitude explosion of an enigmatic space body. And this might not have been realized without Kulik's exploration of the site.

Leonid Kulik's second important discovery during this expedition was a vast zone (8 km across) of trees scorched and devoid of branches, but standing upright like telegraph poles at the center of the radially leveled forest. However, Kulik did not understand the true meaning of this amazing zone of standing trees and explained its existence superficially as caused by a "wave interference." He considered it self-evident that pieces of a meteorite had hit Earth to form the "lunar-like" craters. And although the pattern of the standing trees did appear to him fairly interesting, he thought this of no great importance. Twenty years later, this "fairly interesting" phenomenon led Alexander Kazantsev to the conclusion that the Tunguska space body had exploded in the air, not on hitting the ground.

Kulik's third discovery was to follow. Traces of "unusual burns" were found on both fallen and living trees. "All former vegetation in the hollow and on the neighboring mountains," wrote Kulik, "out to several kilometers, has distinctive traces of a continuous and even burn, which is very different from the traces of a forest fire. These burns have been preserved both on fallen and standing trees, as well as on remains of bushes and moss. They may be seen on the slopes and tops of mountains, in the tundra and on set-apart isles in water-covered swamps. The area showing traces of the burn is several tens of kilometers across." Here Leonid Kulik does deserve praise for his keenness of observation as a true naturalist. Subsequently it became evident that this burn resulted from a powerful light flash during the Tunguska explosion. In the 1960s, having examined the traces of burning, other scientists calculated that the heat radiation from the light flash, in the overall radiation of energy from the explosion, was not less than 10% and perhaps even 25% of the total energy released. The explosion was therefore not only a high-altitude one but, in this respect, rather like a nuclear explosion.

Kulik's discoveries in 1927 were therefore sufficient to understand that the space body that exploded over the taiga in June 1908 could not have been an iron meteorite, although this conclusion was reached only by the great effort of many scientists. And it wasn't just mental effort. When Kulik and his companions had to leave the taiga, their food reserves were running so low that they were tempted to eat their poor horse. "We had provisions just for three to four days, and we were faced with a long trek. Far from being triumphal it was a flight in the literal sense of this word." Although having become noticeably thinner, the members of the expedition (the horse included) reached Vanavara on June 24, and in September both Leonid Kulik and Oswald Guelich returned to Leningrad.

To call Kulik's expedition of 1927 just "successful" would be to underestimate its true significance. It was definitely epoch-making, but somehow the Academy of Sciences was not in a hurry to acknowledge this fact. After all, what did Kulik find? A leveled forest? But that could have been due to a hurricane, something not exactly rare in the taiga! So there were traces of a burn and a forest fire, but no meteor craters, only some holes in the ground! These were not the voices of the uninformed; they were the views of scholars who were familiar both with wind-generated wood falls and with the results of forest fires. The only difference between the critics and Leonid Kulik was that Kulik had visited the place and they had not. And he was sure that the place of the Tunguska meteorite fall was worthy of further investigations, especially as pieces of the meteorite, which could weigh tons, might still be excavated from the ground.

In February 1928, Vernadsky convened a special conference in the Mineralogical Museum on one question only: whether or not the Academy of Sciences should continue the search for the Tunguska meteorite? Opinions at the conference were divided. Some scholars, after studying the photographs taken by Kulik, could not see anything strange or anything needing further investigation. The Academician A. A. Grigoryev, an expert in forestry, suggested that the leveled forest in the "Great Hollow" could have resulted from a forest fire. He did admit, however, that the scale of the event would have had to be extraordinary. The craters at the center of the area of the leveled forest seemed especially doubtful to many at the conference, even to those who generally supported Kulik's work. Nevertheless, they did not rule out the possibility that a large meteorite had fallen in the area in 1908. So the conference resolution was positive: Kulik must go to the taiga once again and finish his work. Either the remnants of the space body would be found or he would find nothing unusual. That was the thinking at the time, but nobody suspected that the unanswered questions about the Tunguska space body would drag on into the twenty-first century.

The Academy of Sciences was then, as always, in straitened circumstances and had to appeal to the government for further funding. The Council of People's Commissars responded favorably so that on April 6, 1928, Kulik was able to leave Leningrad again for the Tunguska taiga with a staff of two people. It's interesting that Kulik took with him not a geologist or an astronomer but a simple enthusiast in the search for the meteorite, a 21-year-old zoologist and hunter Viktor Sytin (1907–1989). Later Sytin, who became a well-known writer, recounted his impressions of this expedition to Alexander Kazantsev, a science fiction author who was to play a sensational role in the Tunguska mystery. Sytin's recollections intrigued the science fiction writer, who began to realize that here was an enigma to explain and that the word "meteorite" was just being used as a convenient label.

On April 25, 1928, the expedition reached Vanavara. There Kulik and Sytin met Nikolay Strukov, a cameraman from *Sovkino* (a state-owned company that controlled the film industry in the USSR from 1924 to 1930), to make a film about the expedition. Kulik hired five local workers, and within a month they had built three *shitiks* (traditional boats). He named them "Comet," "Bolide," and "Meteor." On May 21 with eight in the expedition, they moved downstream to the Podkamennaya Tunguska River and then upstream to the Chamba River, where they hired two extra men to help tow the heavy boats against the flow and the dangerous rapids. On the fifth day, the expedition approached the Burkan mountain range, where the Chamba was rushing down through a narrow gorge. Strukov filmed the expedition surmounting this obstacle where Kulik barely escaped sudden death. Later, Sytin wrote: "The *shitik* was momentarily swamped, turned sideways to the stream, and overturned, and Kulik vanished in the whirlpool... For several seconds, or maybe even minutes, we could not see him. The overturned boat was the only thing that appeared and disappeared amongst the waves and foam... But finally he emerged. We threw him a rope and he clambered on to the bank..."[12] All Kulik said was: "Look here, friends, my spectacles are intact."

Early in June the expedition arrived at "Camp No. 13," built a year before on the Khushmo riverbank. It was a good base, because the distance between the camp and the center of the leveled forest was only a few kilometers. They built a bathhouse and a *labaz* (storehouse on poles: see Figure 3.6). On June 22, the expedition moved closer to their work area – into the "Great Hollow." And near the foot of the Stoykovich Mountain they organized another

FIGURE 3.6. A *labaz* (storehouse on poles) built in the course of the second Kulik expedition (1928) (*Credit*: Dr. Gottlieb Polzer, Lichtentanne, Germany.).

camp. Here they built a log cabin and a second *labaz* and named the place "Meteoritic *zaimka*," a Siberian term for a hunter's house or lodge.

Having finished his filming, Strukov left the expedition with three other workers. Later he made a documentary "To the taiga in search of a meteorite," which contained important material both about the second of Kulik's expeditions in 1928 and about the area of the Tunguska meteorite fall. The rest of the expedition remained to do surveys and prepare magnetometric measurements to try and find the large iron mass of the meteorite that everyone thought was under the ground or in the swamp. They also cleared paths through the taiga to examine the central part of the leveled forest and attempted, without success, to dig up two supposed "meteoritic craters." But as they dug the holes just flooded with subsoil water.

Despite it being summertime, the expedition soon began to feel the shortage of food and vitamins. Their hopes for food from hunting and fishing turned out to be too optimistic, and the explorers had to feed on flour and tea with sugar. There was nothing else and no money left to buy provisions in Vanavara. Sytin and both the remaining workers suffered vitamin deficiency, but Kulik stayed healthy and cheerful. Unfortunately, the measurements for evidence of magnetism in the craters needed to detect meteoritic iron could only be carried out in autumn, when the first frosts would strengthen the soil. So what was to be done? Kulik decided on the risky option of remaining. "We have a food reserve that will last me three months," he told Sytin. "During that time you will reach Moscow and Leningrad, obtain additional funds, and go to Kezhma to arrange for a string of carts to return here for me and our collections."

Kulik's decision to remain on his own was risky, since the taiga even in summer is not completely safe. But even with the food reserves consisting of only flour, tea, and sugar, it proved to be a good decision. Sytin obtained money from the Academy of Sciences and arranged with local Siberian authorities to send a rescue expedition to Kulik. Heading this rescue mission was none other than Innokenty Suslov, the very man who had questioned the Evenks in 1926 about the Tunguska meteorite fall, and he now at last had an opportunity to see with his own eyes where it all happened. On October 20, 1928, they reached Kulik's *zaimka*, and as it was already freezing and snowing they could check for meteoritic iron – mainly in the largest crater that Kulik named "Suslov's crater" after the enthusiastic ethnographer. Alas, no magnetism from such a source was found. But Kulik remained completely unaware of the surprise discovery that this crater would give him the following year.

On October 27 the expedition set out for home as the frost became harder and harder. After two days rest in Vanavara, they journeyed on through snowdrifts in a temperature that was never better than –39°C. When the party arrived at Kezhma on November 6 all were ill, even the iron man Leonid Kulik. Innokenty Suslov had a frost-bitten nose and boils. But after a week's rest these incredible people moved on to the railway station at Taishet from where a fast train – the *Trans-Manchurian Express* connecting Beijing to Moscow – carried them back to civilization.

Soon after arriving back in Leningrad, Kulik started to prepare for the next expedition to Tunguska. It was obvious that a new visit to the "Land of the Dead Forest" must be better organized, or it would fail. On January 2, 1929, at a conference held by the Mineralogical Museum, Kulik read a paper before a large audience on the results of his explorations. He was absolutely certain that the craters in the Tunguska taiga were meteoritic craters, but specialists in the natural life of Siberia disagreed. These are not craters, they said, only natural thermokarst holes. The only way to resolve this disagreement would be to drill holes in several craters until bedrock was reached, but this would need a new expedition.

On January 5, 1929, the Academy of Sciences decided that the new expedition would be sent within the year. Its main aim would be the excavation and drilling of the supposed craters, as well as hydrological investigations of local marshes. The Academy was not slow to act. On February 24, 1929, the third Tunguska expedition left Leningrad and on April 6 it arrived at its place of work. This time it was a well-equipped expedition with 10 well-qualified members, not just a couple of specialists and a few workers. The Academy appointed Evgeny Krinov (see Figure 3.7) as Kulik's deputy. He was then a young astronomer, although after World War II he became a

FIGURE 3.7. Dr. Evgeny Krinov (1906–1984), an eminent meteor specialist, Chairman of the Committee on Meteorites of the USSR Academy of Sciences since 1972 till 1984, a participant of the Great Tunguska expedition of 1929–1930 (*Source*: Zhuravlev, V. K., Rodionov, B. U. (Eds.) *Centenary of the Tunguska Problem: New Approaches*. Moscow: Binom, 2008, p. 24.).

member of the Soviet scientific establishment and a leading specialist in meteoritics. The expedition also had a skilled driller and six young meteorite enthusiasts. They had food for one and a half years, plus hand drills, pumps, spades, crowbars, cameras, measuring instruments, meteorological devices, a theodolite, and chemical reagents. All this equipment and food needed 50 carts to transport it to the taiga.

This Great Expedition lasted 20 months and, of course, included a Siberian winter. Its main aim was to find and dig up that meteorite. And every effort was made to do so. Kulik even prohibited his colleagues from going farther than 3 km from their base, and the exploration of the leveled forest was postponed. First they had to dig the soil, especially in Suslov's crater. The level of water within it exceeded that in the similar nearby depressions, so Kulik decided to drain the water to an adjacent hole. For that they had to dig a trench from Suslov's crater to the adjacent crater. By May 25, 1929, a trench 38 meters long, 1.5 meters wide, and 4 meters deep was finished and water gushed from Suslov's crater into the other depression. At the same time, the upper sphagnum cover, still frozen, sank to the silty bottom of the crater, making it look like a huge bowl. What else could this be, thought Kulik, if not evidence of a meteorite fall?

Alas, while cleaning Suslov's crater from silt and moss, the researchers found near its center the stump of a tree broken near its roots. This was an amazing and shocking discovery. The stump stood in its natural position with its roots penetrating the soil. The discovery was utterly unexpected and destroyed all hope that the crater had been produced by the impact of a meteorite. It was now no more than a hole in the ground.

For Leonid Kulik the discovered stump was a catastrophe. He forbade members of the expedition to take photos (although Krinov did take a photograph secretly) and then ordered the team to drill another borehole on the northern edge of Suslov's depression. But after drilling to 30 m no fragments of a meteorite were found. Kulik then shifted his attention to another promising place, the so-called "Cranberry hole." And until the very end of the expedition's explorations he remained sure that this was a "definite meteorite crater."[13]

So Kulik persisted in his hopeful delusion, although his colleagues who were not so fanatical began to accept that their searches had reached a dead end. On one lucky day, when Kulik had left for Vanavara with a sick worker, Krinov took a long walk through the neighboring area and established that all "meteoritic" crater-like holes were only on low-lying marshy lands. This was one more telling argument against their celestial origin. A swarm of iron meteorites would hardly have preferred to impact only on low-lying land, while ignoring the surrounding mountain slopes.

But Kulik was absolutely deaf to such arguments and insisted on even more digging and drilling. Who knows, he reasoned, perhaps some pieces of the Tunguska meteorite could have fallen at other places of the "Great Hollow"? The best way to verify this idea seemed to be aerial photography, and he eagerly expected the Academy of Sciences to provide an airplane and a photographer. But alas his request was shelved for a whole year, and in 1929 the sky over Tunguska remained empty.

In November 1929, while going from the Great Hollow to Vanavara, Krinov got his feet frostbitten so badly that he left Vanavara for Kezhma, where he spent several months in the hospital. To avoid gangrene, a surgeon amputated a big toe, and in March 1930 he had to quit the expedition. Apart from the health problem, there was also tension between him and Kulik, who considered any doubts about the meteoritic origin of the crater-like holes as a "betrayal." Krinov, however, did not bear a grudge against his chief, and after returning to Leningrad he started to campaign for the requested aerial photography. He convinced the Academy of Sciences to apply for a special plane from *Osoaviakhim* (the so-called Union of Societies of Assistance to Defense and Aviation-Chemical Construction of the USSR, a powerful militarized organization with its own aerodromes, radio clubs, and airplanes that existed in the USSR until World War II). Unfortunately, the plane with Boris Chukhnovsky as the pilot arrived at Kezhma only in July 1930, when it was continuously raining. One day Kulik and Chukhnovsky did take off from Kezhma in the direction of Vanavara only to encounter pouring rain that forced Chukhnovsky to turn back. Taking the aerial photographs of the leveled forest in the Great Hollow had to be postponed indefinitely.

By the autumn of 1930 it became clear that there was no sense in continuing the expedition. Despite it being well organized and equipped, no pieces of the Tunguska meteorite had been found, and in October Kulik returned to Leningrad. His mood was not optimistic. He had lost a battle but did not intend to give up. The unsuccessful searches for meteorite fragments in other holes had led him to a new hypothesis: the huge space body fell in the Southern swamp and exploded there, but the craters were hidden in the waters of this swamp. Pieces of the meteorite, each weighing "several hundreds tons at least" would be there. There was simply no other place. Again and again Kulik tried to convince the academic authorities that a new expedition must be sent to the taiga to search and drill and excavate. And the aerial photography of the region must be done as soon as possible. "It is exceptionally important to photograph this area from a plane," he wrote, "and to create from the photos a large-scale map. This would allow us to understand the nature of the phenomenon much better. There is no other method whose efficiency would be comparable to aerial photography."[14]

But attitudes toward the Tunguska problem had changed – both in society and at the Academy of Sciences. One member of the expedition, Sergey Temnikov, sent a report to the authorities accusing Kulik of incompetence: "He has squandered the people's money, inventing a fantastic meteorite whereas the forest in the Great Hollow was leveled by a hurricane." This was, by the way, not the first and not the last "hypothesis" of this sort. However, leading academics, in particular the president of the Academy of Sciences A. P. Karpinsky, supported Kulik, and Temnikov's report was officially ignored. Temnikov was somewhat too hasty. A few years later this affair might not have ended so easily for Kulik. He might have been accused of "sabotage on the meteoritic front" and joined other exiled scientists in his beloved Siberia, or even further away.

Nevertheless, the Academicians were no longer in a hurry to ask for money from the state budget for Kulik's proposed expeditions. And they were right: it was time to ponder the problem. The picture of the falling space body that had recently looked to be an understandable phenomenon became stranger and stranger, something that Vladimir Vernadsky, who called the Tunguska meteorite an "enigmatic phenomenon," had already realized. It seemed that something important had been missed. At the time there was no accepted theory of crater formation from impacting meteorites, but it was obvious that the vast area of leveled forest testified to the release of an enormous amount of energy whatever the precise nature of the phenomenon. But a meteorite would certainly have left a colossal crater, and no crater existed.

There was a need to just sit down and think, but not for Leonid Kulik. He wanted a tangible stone or piece of iron from space, not a lengthy discussion about abstract questions. For that reason Kulik took almost no part in further theoretical considerations of the problem. He was quite content with the iron meteorite hypothesis that he had accepted at the very beginning of his searches, although he did admit that it might need minor modifications. But certainly, the main impetus to theoretical Tunguska studies came from none other than Leonid Kulik through the discoveries he made himself in the Siberian taiga.

The first major modification of the meteorite hypothesis was that a comet had caused the explosion. This was a reasonable idea since the Solar System has plenty of comets and – as far as we know – only two types of objects can collide with Earth: meteorites and comets. Initially, the Tunguska event was ascribed to a meteorite because of eyewitness reports – and no one knew anything about comets hitting Earth in the past. So a large meteorite provided a ready and acceptable explanation, and even today the world's encyclopedias still describe the Tunguska event as the greatest meteorite impact in recorded history. (One actually carries a photograph showing an alleged piece of that meteorite.) But when the meteoritic model did not match the reported circumstances of the event there seemed to be only one other option: a comet. In one sense, this was not a revolutionary conclusion. Leonid Kulik himself in 1926 thought that the Tunguska meteorite could have been an iron body from a group accompanying the Pons-Winnecke's comet, which could easily be seen in the sky in 1927.[15] This comet, discovered in 1819, was seen in the sky in 1909, fairly soon after the Tunguska event. By the way, on June 26, 1927, it flew past Earth at a distance of only 6 million kilometers – closer than any other comet except one. (Only Lexell's comet in 1770 is known to have approached closer.)

At that time astronomers believed the comet core was probably a conglomerate of stones and dust, or even a simple swarm of meteoroids.[16] So any serious difference between an individual meteorite and a comet seemed difficult to define. However, it was Francis Whipple, then chief astronomer at Kew Observatory in

London, who took the crucial step in 1934 of supposing that the Tunguska meteorite was not just a modest stone – one of a comet's escort – but the comet itself or its nucleus.[17] Unlike Kulik, Whipple thought the cause of the catastrophe was not the Pons-Winnecke's comet but a minor comet that could have been missed by astronomers. As a matter of fact, the same hypothesis was proposed, four years before Whipple, by the American astronomer Harlow Shapley – but in a book, not in a scientific paper.[18] This may be why Shapley's idea went practically unnoticed: scientists prefer their professional journals to books. However, Whipple's hypothesis did offer a reasonable explanation for the puzzling atmospheric phenomena of June 30–July 1, 1908. But his idea did not go far enough. He wrote about a collision of just a comet's core – *consisting of a number of meteorites* – with Earth's surface. This would have left pieces of the comet core and craters at the impact site, but none had been found.

One could probably be sarcastic about Francis Whipple, a theorist who had never visited the Tunguska site. His modification of the Tunguska meteorite model was too limited and his notion of the structure of comets very vague. But this sarcasm would be unfair. Science progresses through the failure of most hypotheses, and if we know more today about the world we live in it is due to former generations of scientists who had to think and work with less knowledge than we enjoy today. Francis Whipple did lay a foundation stone for the model of the Tunguska space body that 30 years later became the favorite of the astronomical community.

In the USSR, Whipple's idea was taken up and strongly supported by Igor Astapovich (1908–1976), an investigator of meteors and meteorites whose book *Meteor Phenomena in the Atmosphere of the Earth* is still considered an authoritative work.[19] In the mid-1930s, he was a young but experienced scientist, and the Tunguska meteorite interested him. When on scientific trips to the basins of the Lena and Angara rivers in the years 1930–1932, he visited 27 places where the Tunguska meteorite had been seen or heard and he questioned witnesses.

So, it was Whipple and Astapovich who almost simultaneously and independently began to study the recorded traces of the Tunguska explosion, which had been made in various parts of the world by seismographs and barographs. And in 1930 Francis Whipple published a paper that used this data to make the first estimate of the magnitude of the Tunguska event. His estimate was 8 kt of TNT. Astapovich in 1933, using almost the same data, arrived at a much higher figure: 25 kt of TNT. Not to be outdone, Whipple revised his calculations and came up with an even higher figure: 50 kt of TNT. At the time the effects of so much TNT were unknown in the real world. Not until an atomic bomb exploded at the Alamogordo Test Range on July 16, 1945, providing the equivalent of 20 kt of TNT, could the effects of such explosive power be seen. A more reliable figure for the Tunguska explosion, calculated by specialists between the 1970s and the 1980s from better data and more precise theories, is 40–50 Mt of TNT. The most powerful hydrogen bomb ever tested on this planet had just this same TNT equivalent – 50 Mt. This explosion took place on October 30, 1961, on the Soviet testing ground of Novaya Zemlya. But in the 1930s the figures obtained looked sufficiently impressive, even though nobody then bothered to measure explosions in kilotons – or still less in megatons.

And what about Leonid Kulik? How did he respond to these findings? He did not respond at all. Certainly, Kulik was still in discussions about the problem of the Tunguska meteorite, but the results of these were only of interest to him as far as they confirmed his own opinion: there was a catastrophic event in the Siberian taiga accompanied by a powerful release of energy. Yes, the results obtained by Whipple and Astapovich strengthened somewhat Kulik's position, but they could hardly be considered crucially important. After his three expeditions, hardly anybody would doubt that "something did fall" in the taiga, even though that "something" had not as yet been excavated. So the skeptics became silent or more cautious when expressing their mistrust. As for Kulik, he understood well that the prospects of further expeditions were uncertain and therefore he temporarily turned to the search for and the examination of other meteorites, enriching the collection of the Mineralogical Museum. Being only slightly interested in theories, he was waiting until there would be a new opportunity to dig the taiga again. But of course Kulik did not forget about the enigmatic Tunguska space body and published articles on this subject from time to time.[20] And he never lost hope that it would become possible to fulfill a long-contemplated plan of taking aerial photographs of the Tunguska site.

Meanwhile, new catastrophic shock waves racked the country: collectivization, industrialization, and, the most terrible of all, the Great Terror of the years 1936–1938. In the 1920s, scientists in the USSR had enjoyed some freedom, but in the Great Terror it was time to stand to attention and be submissive. It is not difficult to understand that in these conditions the Academy of Sciences became less interested in extensive research work in the field of meteoritics. But science still existed and – believe it or not – moved forward. In 1934, by governmental order, the Academy moved to Moscow, closer to the Kremlin. The Mineralogical Museum, including Kulik himself, also moved and for two months, until they obtained a flat in Moscow, Kulik's family lived in his study in the museum, while Kulik slept on his own desk at night.

Soon after the academic institutions arrived in the capital, the Meteorite Department of the Mineralogical Museum was transformed into the Commission on Meteorites. Its academic ranking had definitely risen. Academician Alexander Fersman became Chairman, Vladimir Vernadsky Deputy Chairman, and Leonid Kulik its Learned Secretary. In 1939, the Commission was to become the Committee on Meteorites, headed by Vernadsky, and it would play an important part in postwar investigations of the Tunguska problem.

Eventually Kulik's dream of photographing the Tunguska site from the air seemed likely. And on March 14, 1937, the Presidium of the Academy of Sciences asked for this to be carried out. In May, Kulik arrived at Krasnoyarsk to a city flooded by water from the Yenisey. This delayed him for two months. Only in July when the flood had subsided did a hydroplane equipped with aerial cameras land at Krasnoyarsk. It then took Kulik to Vanavara where, trying to land on the Podkamennaya Tunguska River, the plane crashed. Kulik and his companions survived, but taking aerial photographs was no longer an option, although Kulik visited the Great Hollow before returning to Moscow. His plan to photograph the site had to be postponed yet again.

However, in July 1938 Kulik's persistence and determination were rewarded: a hydroplane was made available to take Kulik and his team to Kezhma, the old Russian village on the Angara River. During the whole of July, photographer S. V. Petrov took pictures that he and Kulik processed, identifying the photos and composing a photographic map. July is perhaps the worst possible month for aerial photography. A riot of vegetation and leaves overshadowed the trunks of the trees felled in 1908. But the results were not bad. A year later, the journal *Reports of the USSR Academy of Sciences* published Kulik's paper: "Data on the Tunguska meteorite for the year 1939." Kulik wrote: "By assembling a mosaic it is possible to determine the initial point from where the main blast wave originated. This center coincides, not surprisingly, with the point that the author determined in 1928 by direct theodolite surveys of the leveled trees. As for additional separate explosions, we can see on the photo assembly two to four such points." (See Figures 3.8 and 3.9.) So Kulik's work showed the structure of the central zone of leveled forest to be very complicated, which meant that the Tunguska explosion had been remarkable for its intricacy. But these important details would only become understandable several decades later.

Regretfully, the priceless negatives of the aerial photographs taken at Tunguska in 1938 (1,500 negatives, each 18 × 18 cm) were burned in 1975 by order of Evgeny Krinov, then Chairman of the Committee on Meteorites. It was done under the pretext that they were a fire hazard, but the truth may have been the active dislike by official meteorite specialists of anything associated with an unyielding enigma. Fortunately, positive imprints were saved thanks to Nikolay Vasilyev, the leader of the Independent Tunguska Exploration Group (ITEG), and they are now at the Russian city of Tomsk, preserved for future studies that might provide new information about the Tunguska space body.

There was another expedition in 1939, the last in which Kulik participated. Its purpose was to link the aerial photographs to points on the ground. It was only moderately successful, but Kulik did not miss the opportunity to thoroughly drill the bed of the Southern swamp. No traces of a meteorite were found. Two years later, on June 22, 1941, Hitler invaded the Soviet Union. Kulik, who was already 57 years old, joined the people's volunteer corps and became a first sergeant of the field engineer company of the first battalion of the 1,312th regiment. The Presidium of the Academy of Sciences attempted to recall Kulik, but he refused to return to the home front. In a letter to his family, dated September 28, 1941, Leonid Kulik wrote: "A bivouac. Tents. Dugouts. The magnificent Milky Way

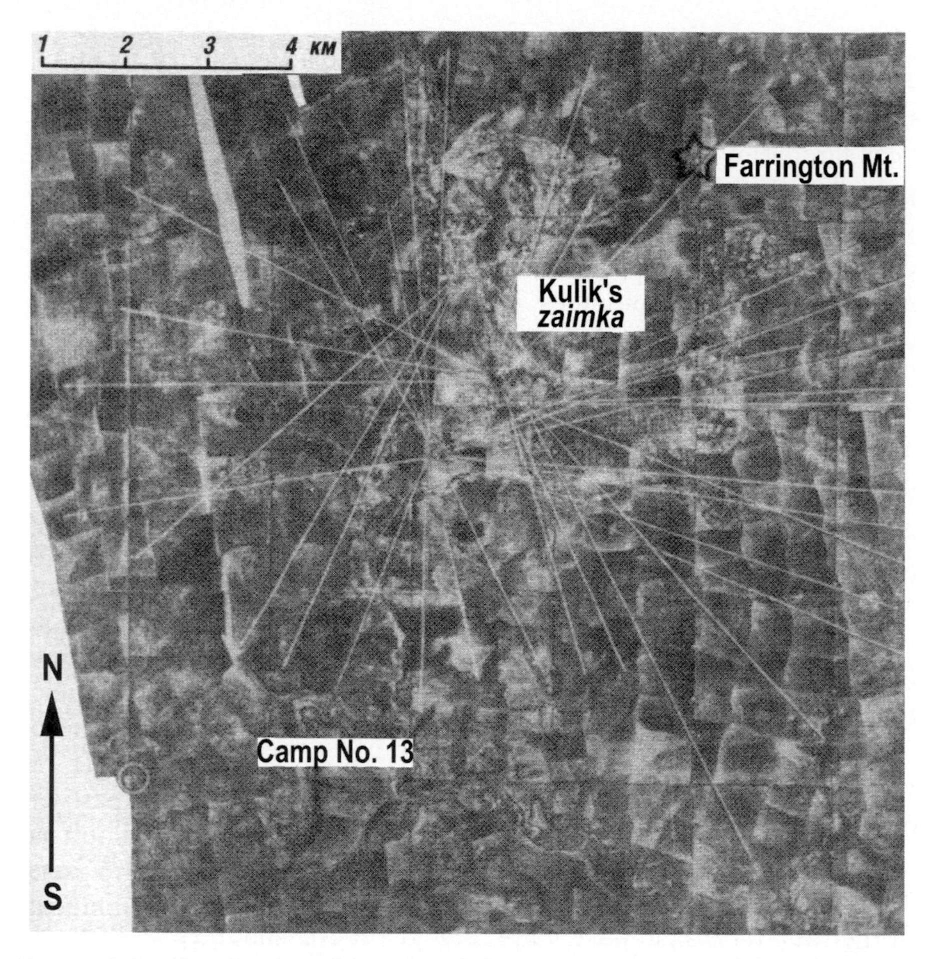

FIGURE 3.8. The photographic map of the epicentral zone of the Tunguska explosion composed by Dr. Leonid Kulik from the aerial photographs taken in July 1938. (*Source*: Krinov, E. L., *The Tunguska Meteorite*. Moscow: Academy of Sciences of the USSR, 1949, p. 155. The whole set of negatives was destroyed in the 1970s by order of Dr. Evgeny Krinov.).

over our heads. A dome of bright lambent jewels covers the Earth, and there flow among this inexpressible beauty the even light of the enormous golden Jupiter, dim leaden Saturn, and the ominous orange-red Mars; the latter leads the way: it rises earlier and stands for a longer time high in the sky, illuminating the lands seized by hurricanes and follies of the war, my poor country among them…"[21]

Eventually there was fighting, poorly armed volunteers against professional Nazi troops. The volunteers were encircled and captured.

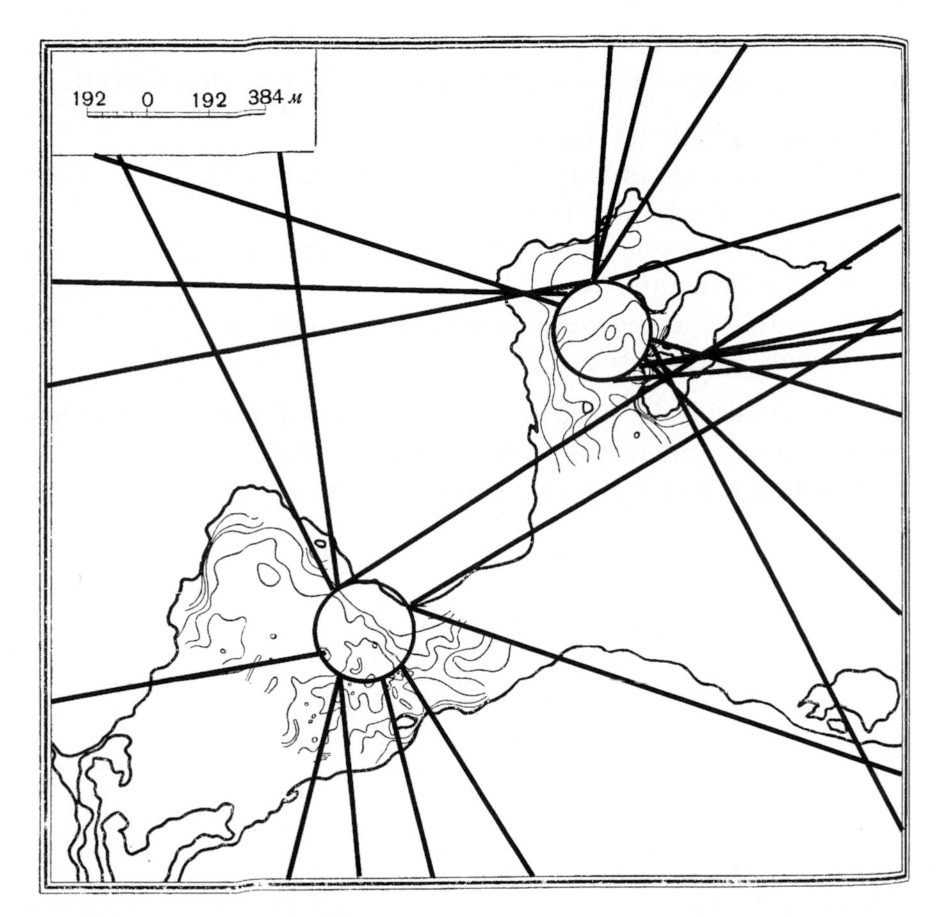

FIGURE 3.9. A drawing of the western half of the Southern swamp with two local epicenters – made by Dr. Leonid Kulik from the photographic map of the epicentral zone. Here two small fragments of the Tunguska space body seem to have exploded (*Source*: Krinov, E. L. *The Tunguska Meteorite*. Moscow: Academy of Sciences of the USSR, 1949, p. 146.).

Kulik was wounded in the leg and became a male nurse in a German concentration camp for Soviet prisoners of war, first in the village of Vskhody and then in the town of Spas-Demensk in the Smolensk Region. It was hellish work, and although his Siberian travels had hardened him he contracted typhus and died on April 14, 1942. By a miracle his grave in the town cemetery has remained intact.

Undeniably, Leonid Kulik's role in the early stages of Tunguska studies was all-important. Were it not for his enthusiasm, which verged on fanaticism, the Tunguska meteorite mystery might have been forgotten forever in the 1920s. Kulik's energies and aspiration for truth overcame his opponents and established the most essential facts about this event. Leonid Kulik made four crucial discoveries:

First, the radially leveled forest.
Second, the zone of branchless "telegraph trees" standing at the center of the leveled forest.
Third, the "unusual burn" covering trees that both perished and survived the catastrophe of 1908.
Fourth, that there were no fragments of a meteorite to be found anywhere at the site.

But ironically the meteorite that Leonid Kulik did not find has become his most important discovery. This is not a play on words. This is a fact. In the next chapter we will have an opportunity to see why.

Notes and References

1. Kulik, L. A. *In Search of the Tunguska Miracle*. Krasnoyarsk: Krasnoyarsky Rabochy, 1927, p. 5 (in Russian).
2. See Svyatsky, D. O. Looking for the debris of a comet. – *Vestnik Znaniya*, 1928, No. 18 (in Russian).
3. Kulik, L. A. *In Search of the Tunguska Miracle*. Krasnoyarsk: Krasnoyarsky Rabochy, 1927 (in Russian).
4. Kulik, L. A. Report of the meteoritic expedition about work carried out from May 19, 1921 till November 29, 1922. – *Herald of the Russian Academy of Sciences*, 1922, Vol. 16, series VI, pp. 391–410 (in Russian).
5. Khatanga is the local name of the upper reaches of the Podkamennaya Tunguska River.
6. See Obruchev, S. V. About the place of the 1908 Great Khatanga meteorite fall. – *Mirovedeniye*, 1925, Vol. 14, No. 1 (in Russian).
7. Voznesensky, A. V. The fall of a meteorite on June 30, 1908, in the upper reaches of the river Khatanga. – *Mirovedeniye*, 1925, Vol. 14, No. 1 (in Russian).
8. Suslov, I. M. Questioning witnesses in 1926 about the Tunguska Catastrophe. – *The Problem of the Tunguska Meteorite*. Vol. 2. Tomsk: University Publishing House, 1967, p. 21 (in Russian); Suslov, I. M. Questioning witnesses in 1926 about the Tunguska Catastrophe. – *RIAP Bulletin*, 2006, Vol. 10, No. 2, p. 17 (in English).
9. Kulik, L. A. *In Search of the Tunguska Miracle*. Krasnoyarsk: Krasnoyarsky Rabochy, 1927, p. 15 (in Russian).

10. By now, this term is used both in a narrow and in a broader sense – either as a designation of the small area (some 8 km across), with the Stoykovich mountain at its center, over which the Tunguska space body exploded, or as the name of the whole region of the Tunguska catastrophe, encircling the whole zone of the leveled forest. In this book, this name is used mainly in the broader sense.

11. Kulik, L. A. *In Search of the Tunguska Miracle*. Krasnoyarsk: Krasnoyarsky Rabochy, 1927, p. 29 (in Russian).
12. Sytin, V. A. *In the Tunguska Taiga*. Leningrad: P. P. Soykin, 1929 (in Russian).
13. Krinov, E. L. *The Tunguska Meteorite*. Moscow: Academy of Sciences of the USSR, 1949 (in Russian).
14. Kulik-Pavsky, V. A. *Life Without Legends: Leonid Alekseyevich Kulik, life story*. Volgograd: Print, 2003 (in Russian).
15. See Kulik, L. A. Meteorites of June 30, 1908, and the crossing of the orbit of the Pons-Winnecke's comet by the Earth. – *Reports of the Academy of Sciences of the USSR*, Series A, 1926, October (in Russian).
16. See, for example: Astapovich, I. S. New investigations of the fall of the Great Siberian meteorite of June 30, 1908. – *Priroda*, 1935, No. 9 (in Russian).
17. See Whipple, F. J. W. On phenomena related to the great Siberian meteor. – *Quarterly Journal of the Royal Meteorological Society*, 1934, Vol. 60, No. 257.
18. See Shapley, H. *Flight from Chaos. A Survey of Material Systems from Atoms to Galaxies.* New York: McGraw-Hill, 1930.
19. See Astapovich, I. S. *Meteor Phenomena in the Atmosphere of the Earth*. Moscow: Fizmatgiz, 1958 (in Russian).
20. See, for example: Kulik, L. A. The 25th anniversary of the Tunguska meteorite. – *Mirovedeniye*, 1933, Vol. 22, No. 2 (in Russian).
21. Kulik-Pavsky, V. A. *Life Without Legends: Leonid Alekseyevich Kulik, Life Story*. Volgograd: Print, 2003 (in Russian).

4. Ideas Become Bizarre

After World War II, the Soviet Union found itself with many problems. Most pressing was the need to rebuild the economy and to develop new weapons. The United States ended the war as the world economic leader, whereas the USSR, which hardly had been an economic giant before the war, had about a third of its national wealth destroyed. Its war casualties reached 27 million. So it was on this foundation that the country had to meet the minimal needs of its citizens while building up its military capability. Naturally enough it made use of German expertise, since Germany caused the war. So several groups of military, science, and intelligence officers were sent to Germany to find and bring back to the Soviet Union plants, machine tools, and high technologies, as well as German scientists and engineers who could help in developing new weaponry in the country. One of those Soviet specialists was Colonel Alexander Petrovich Kazantsev (see Figure 4.1) – the science fiction writer already mentioned – who, in 1945, was chief engineer at a large Soviet research center. At the time he was already the source of several important inventions and had started to write science fiction. Just before the war his first novel, *The Burning Island*, was published.

Alexander Kazantsev was born on September 2, 1906, in the old Russian town of Akmolinsk (now Astana, the capital of Kazakhstan). His paternal grandfather was a merchant millionaire, and his maternal grandfather, a participant in the Polish Uprising of 1863, was sent into exile by the Tsarist government. Before the 1917 revolution, Alexander's father had worked in the family's trading firm, and after the revolution served first in the White Army and then in the Red Army, just as Leonid Kulik had. His mother was a gifted piano player and a music teacher, but Alexander himself graduated at Tomsk Technological Institute in 1930 (not without difficulties because his social origin was not exactly proletarian). This author had the good fortune to become acquainted with

V. Rubtsov, *The Tunguska Mystery*, Astronomers' Universe,
DOI 10.1007/978-0-387-76574-7_4, © Springer Science+Business Media, LLC 2009

FIGURE 4.1. Alexander Kazantsev (1906–2002), an engineer and sci-fi writer, whose hypothesis about the catastrophe of an extraterrestrial starship over Central Siberia gave the main impetus to the Tunguska studies in the USSR in the mid 20th century (*Source*: *The Tunguska Phenomenon: 100 Years of an unsolved mystery*. Krasnoyarsk: Platina, 2007, p. 43.).

Alexander Kazantsev in 1969, and our correspondence, which started as far back as 1963, testifies that he was an outstanding personality. He was not only an inventor and science ficition writer but also a famous chess master, the author of many brilliant endgame studies, and an International Master of chess composition. But what was most important was that he did not fear to think logically, no matter how far this logic might lead him. So it was not an accident that in the mid-1940s Alexander Kazantsev gave a new impetus to the Tunguska studies.

In the spring of 1945, the chief engineer at the All-Union Institute of Electromechanics, Alexander Kazantsev, was given the rank of colonel and appointed the official representative of the State

Committee of Defense (the highest government body in the USSR during World War II) at Vienna. The war was still in progress but it was already time to remove the equipment of Hermann Goering's plants in Styria and to dispatch them to the Soviet Union.[1] Kazantsev completed this task, having survived a serious car accident, and in August 1945 he left Austria for Russia. While driving through Hungary and listening to the radio he heard about Hiroshima and the atomic bomb.

It is worth noting that Kazantsev remembered well Kulik's adventures of the 1920s. In those years he was a student in Tomsk, avidly reading the *Mirovedeniye* and *Vestnik Znaniya* journals, where the circumstances of the Tunguska space body fall were reported, including articles by Viktor Sytin. In 1928, Sytin participated in Kulik's second expedition to Tunguska. And now, while driving back to Moscow, Kazantsev was surprised by the close similarity of the Tunguska and Hiroshima explosions. Having returned to Moscow, he met Sytin, who reassured him that no crater had been found at Tunguska. There had in fact been a zone of standing trees at the center of the area of the fallen forest. Couldn't this mean, thought Kazantsev, that the Tunguska space body exploded in the air and that perhaps the explosion was nuclear? Maybe the meteorite contained a high level of uranium? At that moment, Kazantsev did not think about extraterrestrial spacecraft. He simply tried to bring together the curious aspects of the Tunguska catastrophe into a whole picture. His idea was that the meteorite, or whatever it was, had exploded at altitude over the taiga.

As we know, Leonid Kulik perished in the war, and in January 1945 the other big player in the Tunguska mystery, Academician Vladimir Vernadsky, at 82 years old, also died. So Academician Vasily Fesenkov replaced him as Chairman of the Academic Committee on Meteorites (KMET), and Evgeny Krinov, who was Kulik's deputy in the largest expedition to Tunguska, became its Learned Secretary. The state of affairs in the meteoritic establishment had changed considerably. Vernadsky had been one of the most distinguished geochemists of the twentieth century and a great intellectual, whereas Fesenkov was a noted astronomer and administrator of Soviet science. While Kulik had striven fanatically to discover pieces of the Tunguska meteorite, sweeping away all obstacles from his path, Krinov's approach was different. Even though he had participated in Kulik's searches, he was not at all a fanatic but rather a normal scientist. The science of meteorites interested him much more than the Tunguska meteorite as such. Very probably, when personally visiting the site he understood well (certainly better than Kulik) that hopes of finding any material remnants of the space body were flimsy.

However, in 1945 Evgeny Krinov remained the most authoritative person on the Tunguska problem. Being well aware of this, Alexander Kazantsev planned to contact the scientist, but first decided to meet with other specialists, those who were engaged in nuclear research. After all, he was just a mechanical engineer and science fiction writer, not a physicist or an astronomer, and he wished to make sure that his idea about the nuclear nature of the Tunguska explosion had a rational basis. At the Institute of Physical Problems of the USSR Academy of Sciences, run by the future Nobel Laureate Academician Pyotr Kapitsa (1894–1984), another future Nobel Laureate, Academician Lev Landau (1908–1968), explained to Alexander Kazantsev the principles of atomic explosions. Kazantsev then went to Moscow University to meet a third future Nobel Laureate, Academician Igor Tamm (1895–1971), one of the most prominent Soviet physicists. Tamm had worked in the Soviet nuclear project and later led a group of young physicists, including Andrey Sakharov and Vitaly Ginzburg, who greatly contributed to the creation of Soviet thermonuclear weapons. Both of them, by the way, have also become Nobel laureates.

Kazantsev asked Tamm whether uranium-containing meteorites might exist in outer space, and if so, could one explode like an atomic bomb when entering Earth's atmosphere? No, replied Tamm, it's absolutely impossible. Only atomic bombs can explode as atomic bombs – or at least a similar device built by someone.

If it had been someone other than Alexander Kazantsev talking with Academician Tamm, the whole story might have ended there. Impossible means impossible, and his hypothesis, however attractive, now looked groundless. But Kazantsev was not only an engineer but also a science fiction writer. And as such he thought in a nonstandard way. If the object that vanished in the blaze of a nuclear explosion over the taiga was not natural, it had to be artificial. And since nobody on Earth could have made a device to cause such an explosion in 1908, it had to have been produced by something extraterrestrial.

By that time Kazantsev was going to retire from the army and return to writing. It is therefore hardly surprising that, instead of writing a factual science article, he put his hypothesis into a science fiction short story. The story was entitled *The Explosion*, and it was published in the popular geographical journal *Vokrug Sveta* (*Around the World*) at the beginning of 1946. On the one hand, it was a science fiction story, a literary work with an imagined plot and characters. (There was a black woman claiming to be the sole surviving member of an extraterrestrial expedition, who survived the catastrophe and became a medicine woman in a Tungus tribe.) But on the other hand, the story contained quotations from the papers of Leonid Kulik and real accounts of witnesses of the Tunguska explosion, plus a fairly accurate description of the area of leveled trees. There were some mistakes as well. Kazantsev had overestimated by four times the area of the leveled forest – up to 8,000 km^2 – and underestimated the altitude of the explosion: down to 350 m, approximately that of the explosion at Hiroshima, which was at an altitude of 580 m.

Of course, nobody could have known the exact figures at that time. They were imagined and given simply to fascinate readers. In the story, the superstitious Evenks were wandering through the leveled forests soon after the catastrophe, dreading the wrath of the god of fire and thunder – the dazzling Ogdy. All people who visited the damned place perished from a fearful and unknown disease that covered their internal organs with ulcers. The poor Evenks had become victims of atomic decay from the miniscule remnants of the meteorite scattered in the region of the catastrophe. Yes, remnants of the meteorite. Despite the authoritative explanation of Academician Tamm, Kazantsev proposed that his hypothetical uranium meteorite had caused the explosion. The spaceship hypothesis was mentioned almost in passing at the very end of the story, its author being probably well aware of the potential risk.

It was especially important that Kazantsev plainly stated that the zone of "upright telegraph trees" did testify to the aboveground character of the explosion. He wrote: "Just imagine that: at the very center of the catastrophe, at the swamp that was formerly considered as the main meteoritic crater, where results of the explosion must have been seen most clearly, the forest is still standing upright. To the distance of 30 km all the trees have been felled, but not here. Enormous poles are sticking from the ground... All their branches have been cut by the terrible whirlwind, charring every knot. These trees are so similar to telegraph poles. But why has this dead forest remained upright? Only because the trees were perpendicular to the front of the blast wave. And this could happen only if the explosion did occur at a high altitude above the ground".[2] The lack of a crater and the presence of the "telegraph pole forest" are the main but not the only arguments from Kazantsev for the non-meteoritic nature of the enigmatic space body. His third argument was that the explosion was too powerful for a usual meteorite explosion. His fourth argument was the lack of any meteoritic substances.

Well, perhaps the arguments were rational, but let's not forget that they were set out in a science fiction story, not in a scientific paper, though fantastic stories may sometimes be useful for science, as was so in this case. As it turned out, his readers became fascinated by meteoritics in general and the mysterious event of the Tunguska explosion in particular.

Kazantsev's story was seen by the staff at KMET as a worthwhile piece of science fiction, and Evgeny Krinov accompanied Kazantsev to the Moscow Planetarium to persuade its director Efim Gindin (1898–1966) to start in January 1948 a new teaching program to dramatize the enigma of the Tunguska meteorite. The main role was performed by Felix Zigel, a superb astronomy lecturer, then 26 years old. The plot of this lecture-debate developed dynamically, and its participants came to the conclusion that neither a normal meteorite nor a uranium meteorite could explain the Tunguska explosion and that it could have resulted from an exploding alien spaceship.

In the 1970s, when in Moscow, this author talked with some of the spectators who were at this show. The "first night" of the lecture was attended by leading Soviet astronomers, in particular by Academician Alexander Mikhaylov, Chairman of the Astronomical Council of the USSR's Academy of Sciences and Director of the Pulkovo Observatory. He not only approved Kazantsev's initiative but also congratulated the Moscow Planetarium's team.[3] In the following weeks, the Planetarium's attendance beat all records. Everyone was happy – the author of *The Explosion*, the lecturer, the listeners, and especially Krinov and Fesenkov because they believed it would greatly assist the KMET in popularizing meteoritics.

All this interest could have ended in time. The show would have been removed from the Planetarium's placards and the "Tunguska spaceship" idea would have been forgotten. But both professional astronomers and science amateurs (who were very numerous in the former Soviet Union) were well aware of the results of Kulik's prewar expeditions. They soon saw that Kazantsev's idea was not a simple literary device. It did explain the most unusual aspects of the Tunguska phenomenon. As early as February 1948, Kazantsev's idea became the subject of a serious discussion at a meeting of the Moscow branch of the All-Union Astronomical and Geodetical Society (AAGS).[4] Naturally enough, opinions about Kazantsev's hypothesis were divided, but at the end of the discussion one of the most distinguished Soviet astronomers, Professor Pavel Parenago, said: "I think all of us would agree that it was a space body that fell in 1908 in the Tunguska taiga. What space body it was remains unclear. As for me, I would estimate the chances of it having been an extraterrestrial spaceship as opposed to a usual meteorite as 30–70."[5]

Western specialists at the time would have probably put the chances as no more than 1–99, but the point was that it was a hypothesis worth testing. The idea itself was not mad and could be discussed on a rational level. But the science establishment flew into a rage. It could tolerate a science fiction story, even a staged lecture on the subject, but an attempt to introduce an alien visitation into a scientific hypothesis was not to be tolerated. Why? Nobody knows for sure. Most likely Fesenkov, Krinov, and their colleagues were afraid of the invasion of "dilettantes" into their field of science that dealt with serious astronomical subjects.

In the spring of 1948, there appeared in the newspaper *Moskovsky Komsomolets* (*The Moscow Young Communist Leaguer*) a satirical article entitled "It's strange but a fact" by a Comrade Grekov. Its author expressed his indignation over the "propagation of pseudoscientific figments of imagination" promoted by the Moscow Planetarium. However, soon after this Kazantsev's hypothesis, which the science establishment considered "fantastic," was taken under the protection of the *Komsomolskaya Pravda* (*The Truth of the Young Communist League*) by a noted writer and geographer Nikolay Mikhaylov. The *Komsomolskaya Pravda* ranked higher as a newspaper in the Soviet mass media, but *Moskovsky Komsomolets* did not retreat. Soon it published another article on the subject, authored by three noted scientists: Evgeny Krinov, Kirill Staniukovich, and Vsevolod Fedynsky.[6]

This article was more politically than scientifically oriented. According to its authors, Kazantsev was trying "to propagate under cover of a popular lecture a reactionary cosmological theory of the bourgeois astronomer Edward Arthur Milne" as well as "to intimidate readers with horrible details of explosions of American atomic bombs." These were rather grave accusations at the time – and rather mean as well. In Stalinist Russia in the late 1940s, such accusations were no laughing matter. They could easily bring the accused to the Lubianka cells.

Not all members of the scientific community shared the attitude of these astronomers to Kazantsev and his hypothesis. Several scholars who supported him wrote a letter to *Komsomolskaya Pravda*, but the Komsomol journalists did not dare to publish it, although some excerpts were published in the popular science journal *Tekhnika-Molodyozhi* (*Engineering for Youth*) in the article "On Science Fiction and Wingless Men," written by the reporter Sofya Baratova. The letter defended Kazantsev's hypothesis and was signed by seven professionals in astronomy, including Academician Alexander Mikhaylov and Professor Pavel Parenago, as well as by the faithful associate of Kulik, Victor Sytin.

They wondered on what grounds Krinov, Staniukovich, and Fedynsky had stated that there was no enigma in the Tunguska space body's fall. How could they assert that Leonid Kulik had explained everything when the reality was absolutely different? Also, "such an erroneous approach to this problem precludes the continuation of truly important and – unfortunately – unfinished research that was started by L. A. Kulik."[7]

Academician Alexander Mikhaylov and his colleagues seem to have attempted to return the Tunguska discussion to the field of science free from political overtones. However, the "meteoritic establishment" had taken Kazantsev's encroachment upon their right to decide about the nature of bodies coming from space as an act provoking holy war. Their position was clear: extraterrestrial spaceships belong between the covers of science fiction books; meteorites are a subject for science. When fighting the "dilettantes," meteorite specialists did not mince their words: such terms as "rubbish," "absurd," "antiscientific nonsense" poured from their pens. Soon it became indecent for professional scientists to even consider Kazantsev's hypothesis. In short order, almost all those who defended the hypothesis in the first stage had fallen silent, which was probably a wise move. Few of the scientists involved wished to risk their professional reputations over a spaceship. Academician Mikhaylov hastily went over to the "meteoritic camp" and gave in the summer of 1951 an interview to the popular magazine *Ogonyok* (*A Little Flame*) in which he characterized Kazantsev's hypothesis as fiction.[8] Somehow he also managed to kick the "venal American press" in the same interview because "it had made immediate use of this false hypothesis and had ignored the true scientific facts about the Tunguska event as established by Soviet scientists." He even said that American journalists had written that the Martians also had the atomic bomb ready to invade Earth – probably he was thinking of the Orson Welles radio drama, which caused panic in the streets, and the American press's reaction to it.

Luckily, the indignant newspaper articles denouncing Alexander Kazantsev and his hypothesis as politically harmful did not evoke interest in the Soviet secret police. The State and Party authorities kept mum and left it to the scientists. But in the fall of 1951, after publication of several new anti-Kazantsev articles,[9] the Moscow Planetarium director, Efim Gindin, got sick of constant persecution in the press, and the lecture "The Enigma of the Tunguska Meteorite" was at last closed. The science establishment had achieved a victory.

By that time the KMET people were dealing with another problem that was much more pleasant and promising. A perfectly normal large iron meteorite had hit Earth in full accordance with the rules of meteor science. Like any decent meteorite, it hit the ground and broke into many pieces, which, naturally enough, remained on the site. It was on the clear frosty morning of February 12, 1947, that a bright fiery ball rushed over the Ussury Territory of the Soviet Far East. The duration of its flight was as brief as some ten seconds, but it left behind a long smoky trail that remained in the sky, gradually spreading, for the whole day. Immediately after the bolide disappeared, local people heard loud sounds, like the firing of large-caliber pieces of ordnance, and then a powerful explosion. Witnesses from nearby settlements said that doors in their houses were flung wide open, some window glass broken, and ashes and firebrands thrown out from Russian stoves.

A few days later two pilots were flying at low altitude over the western spurs of the Sikhote-Alin mountain range and saw among the trees a number of fresh craters. To explore them, the Far-Eastern Geological Board sent an expedition from Khabarovsk, which reached the site on February 24, and the geologists found among crushed rocks numerous pieces of an iron meteorite. When the expedition returned to Khabarovsk, a telegram was sent to the Committee on Meteorites of the USSR's Academy of Sciences to report that in the Far East of the country a gigantic iron meteorite had fallen – a very rare event. Later it was named the Sikhote-Alin meteorite. According to the estimates made by Academician Fesenkov, its initial mass, before entering the atmosphere, was about 2,000 metric tons. But almost 95% of this mass vaporized as the meteorite fell through the atmosphere, leaving some 100 tons of first-rate meteoritic iron to reach the ground. The scientists found 106 craters, the largest of them being 28 meters across and 6 meters deep.

Against the background of the Tunguska controversy, which was already flaring up, the Sikhote-Alin cosmic shower proved to be a real heavenly gift to Soviet specialists in meteoritics. The Sikhote-Alin meteorite fall is often compared with that at Tunguska, whereas they are in fact completely different. The former was a normal meteorite fall with craters and iron fragments. The Tunguska event was the explosion of an enigmatic space body with no meteoritic substances or craters. Also, the Tunguska phenomenon produced a noticeable earthquake and the Sikhote-Alin meteorite did not. Even the Vladivostok seismic station, located nearby and possessing very sensitive equipment, did not record any tremor, so the mass of the Tunguska meteorite must have exceeded that of the Sikhote-Alin meteorite by several orders of magnitude. But where is this mass? That is the question.

The results of Sikhote-Alin studies proved to be of prime importance to the world of meteoritics. The collection of meteorites of the USSR's Academy of Sciences, one of the best in Europe, had received many thousands of new meteoritic samples, their total weight being more than 23 tons. At the same time, in the late 1940s and the early 1950s, some attention was still paid by Soviet astronomers to the Tunguska phenomenon as well as to the Sikhote-Alin meteorite. The meteoritic community, despite having become involved in dubious polemics with Alexander Kazantsev and his supporters, continued to work seriously in this direction as well. And Evgeny Krinov, then KMET's Learned Secretary, summed up results of the prewar investigations of the Tunguska event in the brilliant monograph, *The Tunguska Meteorite.*[10]

The main achievement of meteor science after World War II was the theory of crater-forming meteorites, developed in 1946–1947 by Kirill Staniukovich and Vsevolod Fedynsky.[11] Generally speaking, it was always evident that a meteorite moving at a great speed and striking land would most likely vaporize. Thus, in Kazantsev's short story "The Explosion," written in 1945, a supporter of the meteoritic model of the Tunguska phenomenon explains how the taiga was leveled: "The meteorite that flew at a great cosmic velocity hit the ground, and all its kinetic energy was transformed into heat. Hence the explosion."[12] It was Staniukovich and Fedynsky who provided the mathematical support for this conclusion. They showed that if a meteoritic body is moving faster than 5 km/s *just before its impact*, then, immediately after the meteorite strikes Earth's surface, shock waves spread through both the surface material and the meteorite itself. And the meteorite is vaporized completely by the released energy. The shock wave inside the ground projects material upward and outward from the point of impact, thus forming a crater – and no remnants of the meteorite are preserved on the site. But this occurs only if the meteorite's final velocity is really great; otherwise its fragments may be found (as happened at the Sikhote-Alin mountain range).

Evgeny Krinov immediately attempted to apply this theory to the Tunguska problem. He believed it could explain all phenomena that had accompanied this event.[13] Recall that the enormous magnitude of the Tunguska explosion was one of Kazantsev's arguments in favor of the spaceship hypothesis. Kazantsev believed the explosion was "too powerful" for a normal meteorite, but research has shown that an iron meteorite hitting the land could have produced a huge amount of energy without leaving fragments. However, the problem is that a very large crater would have been formed – and Krinov himself, having spent almost a year at Tunguska, had seen no crater.

Of course, the KMET should have sent a new expedition to Tunguska to try to find a drowned crater, if not the vaporized meteorite itself, thus putting an end to Kazantsev's fantastic invention. The KMET people did think about this, but the Sikhote-Alin meteorite fall had grabbed their attention. However, in the summer of 1953, the geochemist Kirill Florensky (see Figure 4.2; a son of the great Russian theologian and philosopher Pavel Florensky, who had been shot in a gulag in 1937) found himself at Tunguska, when exploring gas fields in central Siberia. Evgeny Krinov asked Kirill Florensky to look around and inform the KMET if anything had changed at the Tunguska site during the past 14 years since Kulik's last expedition. He wanted to know if a new expedition would meet with any appreciable difficulties if sent to Tunguska. Also – the

FIGURE 4.2. Dr. Kirill Florensky (1915–1982), a Soviet geochemist and planetologist, a pupil of Academician Vladimir Vernadsky, who headed several Tunguska expeditions organized by the USSR Academy of Sciences (*Source*: Bronshten, V. A. *The Tunguska Meteorite: History of Investigations*. Moscow: A. D. Selyanov, 2000, p. 108.).

question of prime importance – would Florensky look for a meteoritic crater? The geochemist did look, visiting Kulik's *zaimka* and also making a reconnaissance flight over the area of the leveled forest. Florensky made sure that the felled trees were still clearly visible, despite the young growth, but he could find no trace of a crater. His main conclusion was that a new expedition could reach the place with relative ease.[14]

Nevertheless, the next four years passed in vacillations – whether or not such an expedition would justify the expense. Then in July 1957 Alexander Yavnel, a KMET scientist, discovered meteoritic iron in Kulik's Tunguska samples. KMET possessed 89 samples of soil brought back by Leonid Kulik from Tunguska and had kept them in cardboard boxes with tightly closed lids. They had been discovered only by chance when the KMET people were sorting out their archives. Since the most probable place for the fall of the Tunguska space body was the Southern swamp, Yavnel selected 13 samples from that area. Each sample had been ground and a strong magnet had extracted magnetic iron, which was examined under a microscope. The following components were found:

1. Crystals of magnetite.
2. Metallic particles of silver-white color only several tenths of a millimeter long.
3. Oxidized metallic particles with slightly fused surfaces and edges. Usually, they were flat and acute-angled, or looked like bars of a few millimeters in length.
4. Bright black spherules consisting of magnetite, with a diameter of 30–60 microns. There was also a spherule of silver-white color.

The spectral analyses showed that the metallic and oxidized particles consisted of nickelous iron. They were checked at the Institute of Geochemistry and Analytical Chemistry and found to contain 10.5% of nickel. This surprising result seemed to indicate that the Tunguska event had been due to a natural iron meteorite.

"One can say with a fair degree of confidence," Yavnel concluded, "that we possess here the substance of the Tunguska meteorite, and it strongly suggests that it was an enormous mass of iron."[15]

Yavnel sent his paper to *two* scholarly periodicals: *Geokhimiya* (*Geochemistry*) and *Astronomichesky Zhurnal* (*Astronomical*

Journal),[16] and it was soon published in both journals – which was unusual. Papers could wait a year or more for publication. Since the Tunguska polemics were mostly carried out in the popular press, Evgeny Krinov at KMET also invited two well-informed journalists to share the sensational news. Their article, "The Tunguska Meteorite Has Been Found," was soon published in the popular science journal *Znaniye-Sila* (*Knowledge is Power*).[17] The article informed readers that the enigma of the Tunguska meteorite had been solved. It was no spacecraft but a normal piece of cosmic iron. The particles discovered by Alexander Yavnel testified to this. The same news was also published in an article by Yavnel and Krinov in *Komsomolskaya Pravda* (*The Truth of the Young Communist League*).[18] Alexander Kazantsev and other enthusiasts of the spaceship hypothesis were taken aback. Some in despair suggested that the shell of the alien spaceship could have been made of nickelous iron, but KMET specialists kindly explained that this was sheer nonsense. Anyway, the Academy of Sciences decided that an expedition must be sent to Tunguska, to provide a final answer to the question.

Yavnel's discovery, however, was not the only reason for this decision: the jubilee of the Tunguska event was approaching. Half a century had passed since the enigmatic explosion in this remote corner of Siberia; now it was time to solve the mystery. Besides, the first *Sputinik* was launched in 1957, and the spiritual atmosphere in the country was, so to say, space-oriented, making Kazantsev's hypothesis very popular among the young scientific and technical intelligentsia. This worried the KMET people. But then it only remained to go to Tunguska to find there particles similar to those discovered by Alexander Yavnel, preferably in the meteorite crater, and the question would be closed forever.

In the 1950s, specialists in meteoritics stubbornly refused to believe that the Tunguska space body had exploded in the air. Nobody at KMET suspected that there could be neither meteor particles nor meteoritic craters in the taiga – with a probable exception of the experienced but tight-lipped Evgeny Krinov.

In 1958, Kirill Florensky, no novice in the taiga, was appointed to lead the new academic expedition. Apart from him, the team, consisting of 11 people, left by train and then plane for Vanavara's new airport. The whole population of this settlement, closest to the Tunguska explosion site, was then about a thousand. It was one of three district centers of the Evenk Autonomous Region (or Evenkya). The region did, however, remain very sparsely populated: in an area of about 800,000 km^2 there were only 12,000 inhabitants, less than half of them Evenks or Tungus.[19]

The new KMET expedition possessed precise maps of the Tunguska region that Kulik had lacked. Their itinerary was also different. On June 3, 1958, they left Moscow by train for Krasnoyarsk, from where they went by plane to Vanavara, where the local authorities provided the scientists with 40 deer needed for the last stage of their journey. The expedition reached Kulik's *zaimka* on June 27. Three days later they celebrated the 50th anniversary of the enigmatic event that had occurred at that very place. As participants of the expedition later recalled, they marked the occasion with a special bottle of champagne.

According to its final report, the aims of the expedition were (1) search for the crater, (2) search for meteoritic substances, (3) exploration of the leveled forest, and (4) evaluation of further research prospects. The main problem and the main research target was in fact the crater – more than the substance of the meteorite. The problem of the remains of the meteorite appeared to have been successfully solved by Yavnel a year earlier, so that control tests on the site seemed nothing but a formality. But the lack of any crater still made the KMET people nervous. If a crater existed, then the explosion occurred on the ground, and the academic position was correct. If not, then the explosion must have occurred in the air. That is why the expedition had to first examine the Southern swamp – since it was the only possible location of the hypothetical crater – to look for any signs of explosion-related alterations in its bed. Their main concern was to answer this question, but no signs of any meteoritic crater were found. As they reported: "We were unable to find traces of a ground explosion. All members of the expedition have agreed that the Southern swamp could not be the place where the explosion happened that leveled the forest around."[20]

The second task in order of importance was to take soil samples and test them for nickel as a sign of the presence of nickelous meteoritic iron. Fesenkov and Krinov assumed that the expedition would find the dispersed substance of the Tunguska meteorite and be able to determine the area of its highest concentration, indicating the very place where the meteorite had fallen. This was not to be.

Florensky and his colleagues did find in Kulik's *zaimka* samples of soil that had been left there by Kulik himself. A year before, in similar samples, Alexander Yavnel had found meteoritic iron. So Kulik's samples were immediately analyzed. Alas, there was no meteoritic iron in them. The expedition scientists then became even more circumspect and started a very accurate and systematic gathering of samples from the Tunguska soil. Almost every sample contained some small quantities of iron, but never any nickel. But meteoritic iron contains a lot of nickel. So, there was iron at Tunguska – but not meteoritic iron. True, there were in the soil some microscopic silicate and magnetite spherules that could have been of space origin. But these spherules did not differ in composition and amount from the usual space dust that is regularly falling on Earth.

The expedition brought to Moscow almost a hundred new samples of the Tunguska soil, as well as 50 of Kulik's samples that had been kept at his *zaimka*. And these were carefully analyzed with up-to-date equipment – for the year 1958. There were no signs of meteoritic iron in the samples. The content of meteoritic dust corresponded well with usual fluctuations of the background fall of space dust. So the academic expedition had failed to solve the two primary research tasks. Its members could not establish the meteoritic nature of the Tunguska space body, but this "failure," as it turned out, proved to be a great success – the work of the expedition demonstrated that the iron meteorite hypothesis should be rejected.

Of course, having no crater and no meteoritic iron was hardly sufficient to compose a substantial scientific report. Luckily enough, however, the third direction of research – the examination of the leveled forest – proved to be more informative and its results rightly still hold a prominent place in the final report. True, the expedition was unable to determine the borders of the leveled wood with sufficient accuracy, it being just too small for this task. But the expedition collected important data about the felled trees. There were six types of damage recorded that would greatly help in compiling a detailed map of the leveled wood. Making such a map was very reasonably listed under number one in the plan of future investigations, but it was not the Committee on Meteorites that subsequently implemented this important project. A few years later the map was composed by members of the Independent Tunguska Exploration Group (ITEG).

Florensky and his colleagues paid great attention to the "telegraphnik" – the central zone of standing trees. Naturally enough, many of the branchless "telegraph trees" had by that time fallen down in high winds and were lying chaotically. Having crossed this zone several times, members of the expedition realized that it was asymmetrical in relation to the borders of the leveled wood area. This meant that the blast wave had also been asymmetrical.[21] There seemed to be in the Tunguska taiga no usual ellipse of dispersion typical for meteorite showers. The zone of leveled forest was oddly complicated.

The expedition also tried to solve the problem of the "unusual burn," which, according to Leonid Kulik, had evenly covered vegetation in the Great Hollow for many kilometers across. This burn had been very different from the traces of a usual forest fire. Generally, they did not doubt the real existence of this phenomenon, described by the Tunguska pioneer himself, but they were unable to discover its traces and therefore decided that the evidence had already disappeared. Subsequently it turned out that some traces of the anomalous burn persisted but could not be easily found. Analysis of these traces of burning has even formed a separate direction for Tunguska studies. But in 1958 this subject encountered a problem when geologist Boris Vronsky found two old larches in the Southern swamp that had safely survived the Tunguska catastrophe. These were more than 50 years old, but both trees were alive, healthy, and not even burned. One was cut down, and the scientists determined its exact age from the annual rings. It was 108 years old. That two robust trees still existed on the swamp that had been considered a probable meteoritic crater demonstrated that the swamp could not be a crater. At the same time this fact seemed to testify no less convincingly against the nuclear hypothesis. How could the larches have survived an atomic explosion at its epicenter without any burns? Impossible!

After the discovery of the larches, the problem of the anomalous burns lost its topicality for the academic expedition. Its chief decided that there could not have been a powerful light flash at Tunguska. Today, however, there is reason to believe that the undamaged larches on the surface of the Southern swamp may be interpreted differently – as evidence of the uneven character of this light flash. But the flash itself had been powerful indeed; this was subsequently proven by specialists who examined the traces of the light burn.

The accelerated growth of the forest on the territory affected by the Tunguska explosion was another important – and unexpected – discovery made by Florensky's expedition of 1958.[22] Unusually wide tree rings (up to 9 mm wide) were found at the central part of the leveled forest, both in trees that had grown after the explosion and in trees that had survived the explosion. Before the Tunguska catastrophe, the average width of the annual rings was only 0.2–1.0 mm.[23] At first, this effect appeared understandable because due to the explosion the taiga in this region became thinned out and the soil enriched with ash (which served as a fertilizer), which must have led to better growth of all the trees. But this simplistic explanation was subsequently rejected, and the accelerated growth of the forest is now considered as another enigma of the Tunguska phenomenon.

Having returned to Moscow in October 1958 and reviewed the findings of the expedition, the scientists arrived at two important conclusions. First, there was definitely no meteoritic iron in the soils of the Tunguska region, which meant that Yavnel's result was erroneous. Most likely, Kulik's samples that were kept at KMET's building became contaminated when other meteorites (such as fragments of the Sikhote-Alin meteorite) were sawed during research. At present it is hard to say whether this was so, but in any case Alexander Yavnel's mistake proved to be another happy one in the history of the Tunguska problem. Were it not for Yavnel, the academic expedition would not have been sent to Tunguska in 1958, neither, most probably, in the following years.

Having evaluated the collected data, the members of the expedition wrote: "The absence of large deteriorations in the central zone of the leveled forest – that is, on the Southern swamp, as well as the lack of noticeable meteoritic craters and the presence of the 'zone of indifference' in the center of the catastrophe make it possible to suppose that the shock wave of the Tunguska explosion was moving in this region mainly in a downward direction, its center being located high up."[24]

One translation of this text from its scholarly jargon into a clear English is: *The Tunguska space body exploded at a great altitude in the air, and not when hitting the ground.*

A more general conclusion, having significance for the whole science of meteoritics, should have been: "It would be premature to consider the Tunguska meteorite as a typical crater-forming meteorite. The meteoritic theory must be supplemented with a case when vast ground devastation occurs without forming a crater on Earth's surface."[25]

Somehow, Alexander Kazantsev was not mentioned in the final report of the expedition, yet it was Kazantsev who had predicted the two important facts: that on the site of the explosion there would be no meteoritic substance and that it would be proved that the Tunguska space body had exploded in the air. And he did this by using the "spaceship model," however fantastic it may have seemed. Certainly, in his prediction, Kazantsev leaned upon the results of Kulik's expeditions, but the key thing was his ability to look at them from a different theoretical standpoint. Supporters of the meteoritic hypothesis, who had been persistently defending their model of the Tunguska phenomenon for more than 10 years, now had to look for an acceptable explanation of these two facts – alas in retrospect. In other words, the spaceship model took the lead in Tunguska studies.

This is why after the academic expedition of 1958, its participants – and first of all Kirill Florensky – were so perplexed. Everything looked predictable before the trip: they left for the taiga to find the crater and nickelous iron that would have confirmed the normal meteoritic model. But now they had no crater or meteoritic iron – and it also turned out that the "meteorite" must have exploded in the air. Not a pleasant situation for them. But they were scientists and used to dealing with facts. Even if they thought Kazantsev's hypothesis nonsense, they could not dismiss the new evidence from Tunguska. The "evil spirit" of the enigmatic space body had not vanished into thin air, so a scientific explanation had to be looked for. Being rather confused by his own findings, Kirill Florensky sent some samples taken at Tunguska to the Institute of Geochemistry and Analytical Chemistry of the USSR's Academy of Sciences and asked them to check for radioactivity. Taking into account that KMET considered any attempt to investigate radioactive contamination in the Tunguska region as pseudoscience, it was a bold step. The academic chemists, however, discovered no traces of increased radioactivity, and this question was closed – at least temporarily.

Early in the autumn of 1959 the Moscow Institute of Physical Problems held a workshop on the Tunguska event. Mikhail Tsikulin and Vladimir Rodionov contributed the main paper. These scientific workers of the Institute of Chemical Physics of the USSR's Academy of Sciences[26] suggested that the forest devastation in the Tunguska taiga had been caused by the ballistic shock wave that had accompanied the meteorite flying in the atmosphere and had hit the ground after the meteorite had been disrupted by the forces of air resistance.[27] Of course, this model also faced the same old question: where were the remains of the meteorite?

The fact is, however, that every big scientific problem should be approached in stages. Specialists in ballistics had first to settle the main issue of how a piece of iron from space could fell such an enormous number of trees without touching Earth's surface. To test their hypothesis, Tsikulin and Rodionov performed a series of modeling experiments. In a blasting chamber they placed a thick layer of soil, sticking into it a number of bits of wire to represent trees. Over these "trees" the physicists put a detonating cord with an amplifying charge at its end. The blast wave from the detonating cord served as a model of the ballistic shock wave, propagating from a space body flying in the atmosphere. Tsikulin and Rodionov assumed that the meteorite exploded at an altitude of 100–500 m (apparently using the figures proposed in Kazantsev's short story "The Explosion"). The energy then released would have been 10 Mt of TNT, but the altitude was definitely underestimated. More importantly, in 1959 the true shape of the area of leveled wood remained unknown to the investigators. Evgeny Krinov, who spoke at the workshop after Tsikulin and Rodionov, was still doubtful of the overground character of the Tunguska explosion and severely criticized their report. In time, though, his opinion changed.

Incidentally, Alexander Kazantsev attended the workshop and was even allowed to speak. Physicists, as a rule, were ready to discuss his "spaceship hypothesis" sympathetically, as distinct from meteor specialists who would not have let him through the door of a meteoritic conference. But in this case Academician Pyotr Kapitsa, Director of the Institute of Physical Problems, himself decided who could or could not be invited.

As for the chief of the academic expedition, Kirill Florensky, he generally accepted the ballistic model of Tsikulin and Rodionov, even though stating in some articles that the hypothesis of a crater-forming meteorite had not yet been disproved. At the same time, he was not fully satisfied with the purely "ballistic" approach to the Tunguska event. Having twice visited the site, Florensky felt that the forest could not have been leveled just by the meteorite's "energy of motion." There must also have been an explosion, such as a violent release of energy from a chemical or nuclear reaction, in the substance of the space body. But he wouldn't consider a nuclear reaction, so it only remained necessary to modify the "hypothesis of a ballistic shock wave" by supplementing it with some "chemistry." According to Florensky, the Tunguska meteorite, being a natural space body, could have consisted of substances that could have exploded when mixing with atmospheric oxygen. The meteor specialists, however, ignored Florensky's idea, and it was only much later, after his death, that it was noticed and developed by other researchers.

The "purely ballistic" approach to the Tunguska problem attracted the meteor specialists, first of all by its simplicity. Yet some discrepancies with the facts were noticeable. The trajectory of the Tunguska meteorite was gently sloping – all Tunguska investigators shared this opinion. However imprecise the eyewitnesses' accounts might be, they were sufficient to come to that important conclusion. Meteors begin to emit light at an altitude of 130 km or lower. Even if the most distant points where the Tunguska bolide was seen were about 800 km from the place of its explosion (and there were more distant observations), then the slope of its path could not have exceeded 17°. But the experiments of Tsikulin and Rodionov showed that a slope of 30° was needed to reach an acceptable correspondence between the model and the real picture. It was a new enigma that had to be resolved. Generally speaking, this result was self-evident: to fell trees strictly radially, the ballistic shock wave would have had to move in a very steep path. If it had moved flatly there would have been a long belt of fallen trees shaped like a herring bone.

Florensky's "chemical explosion" looked too exotic for meteorite specialists. So they started searching yet again for an acceptable theory to explain the undeniable fact of the radial character of the leveled forest. Such a theory had to combine two main traits. First it had to be a natural cosmic body that had exploded (a meteorite or a comet, but definitely not a spaceship). Second, this body had to produce not only a ballistic shock wave but a vast blast of energy as well. The strictly radial character of the leveled forest testified to the fact that the space body had definitely *exploded*, not simply collapsed in the atmosphere to liberate a ballistic shock wave that hit the taiga. It was therefore necessary to find a mechanism for *a natural overground explosion in a natural space body*.

Physicist and astronomer Kirill Staniukovich, with his colleague Valery Shalimov, developed this acceptable mechanism.[28] There exists an equation for the heat balance of a meteorite flying in the atmosphere. When moving through the air, a space body gets hot because it's gaining more heat than it's losing. According to the equation, at a certain altitude (for iron meteorites at about 18 km) these two processes become balanced, and the meteorite stops heating up. Instead, it starts getting cooler while simultaneously slowing down, so that it falls on the ground moving at a relatively lower speed. For a stone meteorite the picture is practically the same. But for a lump of ice it's different. Such a lump with a diameter of, say, 10 m, moving at the velocity 60 km/s, heats up very intensely. At an altitude of 50 km the heat supply exceeds 10 times what is being lost, and the space body starts to vaporize very actively, a process that rapidly becomes highly violent. This is the so-called "thermal explosion," which might have explained peculiar aspects of the Tunguska catastrophe (see Figure 4.3).

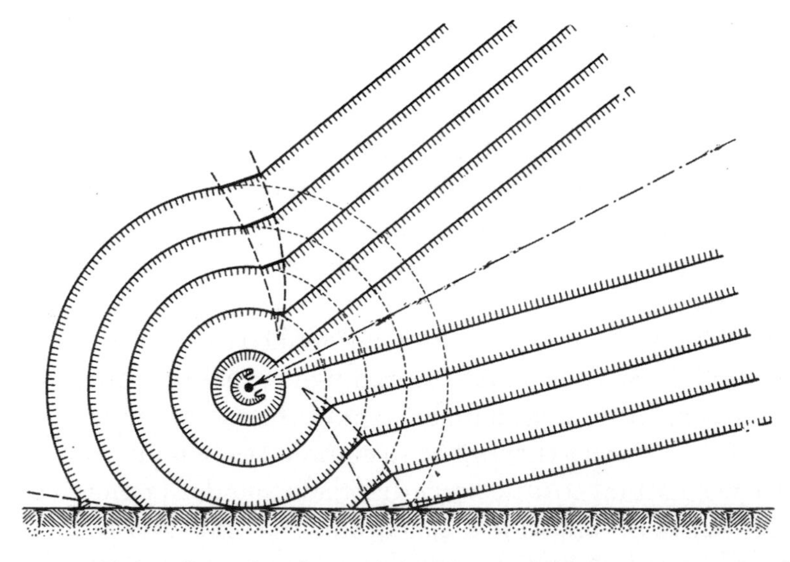

FIGURE 4.3. This is how the thermal explosion of the Tunguska space body must have looked, according to the theory of Dr. Kirill Staniukovich and Dr. Valery Shalimov.

What else was needed to be able to apply the model of Staniukovich and Shalimov to the Tunguska phenomenon? It was ice, no matter whether it be the usual watery ice or frozen gases. Neither stony nor iron meteorites possess properties that would make possible a "thermal explosion." But the icy core of a comet does possess them.

By the 1950s the old model of the comet core as a conglomerate of stones and dust with a small amount of ice (the so-called "flying sandbank" model proposed by the famous English astronomer Richard Proctor in the nineteenth century) passed out of favor. In 1951 the noted American astronomer Fred Whipple developed a new model for the comet core, which much better corresponded to the observational data. In the popular press this model got the name of "dirty snowball," although Whipple himself preferred to call it "the model of icy conglomerates." According to this model, the comet core consists of about one quarter dust, stones, and iron bodies and three quarters ice. And this ice is a mixture of frozen water and frozen gases, such as methane, ammonia, carbon monoxide, and carbon dioxide. Although at one time the comet core was thought to be "stones with some ice," it was now "ice with some stones and iron." Lately, though, specialists in cometary astronomy have started to think that the share of hard substances in comet cores is greater. So we now have the "icy dirtball" hypothesis. The Solar System appears to have *two* types of comets: dirty snowballs and icy dirtballs. And perhaps there are more types we don't know about.

The new stage of the cometary approach to the Tunguska problem is usually associated with the Chairman of KMET Academician Fesenkov. But in fact it was Evgeny Krinov who in 1960 reanimated and substantially revised the "old" cometary hypothesis of the Tunguska space body's origin that had been suggested early in the 1930s by *British* meteorologist *Francis* Whipple. Two years before, Krinov rejected the very possibility that the Tunguska space body could have exploded in the atmosphere and not when striking the ground. Now he wrote: "It comes as no surprise that there is no crater in the area of the meteorite fall, for it exploded in the air".[29] Krinov concluded that the lack of any substance is no wonder either because it was a comet core consisting of watery ice and frozen gases that produced the Tunguska event.

However, somehow this explanation of the Tunguska phenomenon became associated not with Evgeny Krinov, the noted specialist in meteorites, but with the name of his academic boss Vasily Fesenkov. Whether or not Fesenkov was thinking over the possible cometary nature of the Tunguska space body independently of Krinov remains unknown, but his first paper on this subject appeared in the scientific press more than half a year after Krinov's article in *Priroda*. And it was by both Fesenkov and Krinov.[30] But as for the comet's core, Fesenkov still believed it consisted of "very compact dust clouds several kilometers in diameter."

Even a year later, Fesenkov was still vacillating between the "flying sandbank" and "dirty snowball" models of comet cores. He emphasized that if the "dirty snowball" model is correct, then no debris of the Tunguska comet could be ever found. Equally, if the comet core resembled a "flying sandbank," then a swarm of small meteoroids would have been scattered over an enormous territory. At best, he said, we could hope to discover some microscopic spherules that formed from the fused and dispersed cometary substance.[31]

Combining the theory of the heat explosion with a new cometary model of the Tunguska space body proved to be a great achievement for meteor specialists. The overground explosion of the space body had been acknowledged and theoretically explained. And according to this theory, the forest had been leveled not only by the ballistic shock wave but also by a blast, while the lack of cosmic substances on the site of the explosion became explicable. Frozen water and gases (the main components of the comet core, according to Fred Whipple's theory) vaporized, whereas its stony and iron components have dispersed in the atmosphere, slightly contaminating the Tunguska soil.

Of course, this solution somewhat resembled making the theory fit the data. But why not? In science such methods of finding correct solutions are not forbidden. But whether or not the new version of the cometary hypothesis could be taken as the final solution of the Tunguska mystery remained unclear. The meteor scientists wanted this, but after Yavnel's fiasco they became more cautious.

The framework of the cometary/meteoritic approach to the Tunguska problem resulted from many distinguished specialists studying the problem. Using a high level of mathematics they rigorously analyzed the complicated processes going on when an iron, stony, or icy body is flying through the air. These specialists gave lectures at conferences and published scholarly monographs and papers in scientific periodicals. The results obtained contributed to a better understanding of such processes, helping, in particular, to create manned orbital spacecraft and warheads for intercontinental missiles.

The KMET need not have feared Kazantsev's spaceship. "Pseudoscientific sensations" in the Soviet Union had no chance of survival. Even so, participants of meteoritic conferences and symposia did not forget to pass resolutions condemning Kazantsev's ideas as "antiscientific lies" and "the lightheaded hunt for sensations." These resolutions were regularly sent to high officials of the Union of Soviet Writers, together with severe demands to forbid Alexander Kazantsev from writing about the Tunguska meteorite. The future promised to be serene for KMET. It did not, however, keep its promise. Kazantsev's hypothesis, although suggested by a nonprofessional, caused Alexey Zolotov and a large group of Siberian scientists to start their own investigations in the taiga. In the next chapter we will see how crucially this changed the atmosphere of Tunguska studies.

Notes and References

1. For details see Kazantsev, A. P. *Dotted Recollections*. In: Kazantsev, A. P. *The Ice is Returning*. Moscow: Molodaya Gvardiya, 1981 (in Russian).
2. Kazantsev, A. The Explosion. – *Vokrug Sveta*, 1946, No. 1, p. 43 (in Russian).
3. See Zhuravlev, V. K., Zigel, F. Y. *The Tunguska Miracle: History of Investigations of the Tunguska Meteorite*. Ekaterinburg: Basko, 1998, p. 23 (in Russian).
4. The AAGS was an academic organization of the USSR, carrying out research in the fields of astronomy, geodesy, and cartography, and its members were both professional astronomers and amateurs.
5. Zhuravlev, V. K., Zigel, F. Y. op cit., p. 24.
6. Staniukovich, Kirill Petrovich, 1919–1989, an astronomer and specialist in gas dynamics, professor, doctor of physics and mathematics. Together with Vsevolod Fedynsky, he has developed the theory of crater-forming meteorites. Fedynsky Vsevolod Vladimirovich, 1908–1978, specialist in meteoritics, corresponding member of the Academy of Sciences of the USSR.

7. Baratova, S. On science fiction and wingless men. – *Tekhnika-Molodyozhi*, 1948, No. 9, p. 27 (in Russian).
8. See Mikhaylov, A. A. The Enigma of the Tunguska meteorite is solved. – *Ogonyok*, 1951, No. 24 (in Russian).
9. See, for example, Fesenkov, V. G., Mikhaylov, A. A., Krinov, E. L., Staniukovich, K. P., Fedynsky, V. V. On the Tunguska meteorite. – *Nauka i Zhizn*, 1951, No. 9 (in Russian); Fesenkov, V. G., Krinov, E. L. The Tunguska meteorite or... a "Martian spaceship"? – *Literaturnaya Gazeta*, 1951, August 4 (in Russian).
10. Krinov, E. L. *The Tunguska Meteorite*. Moscow: Academy of Sciences of the USSR, 1949 (in Russian).
11. See Staniukovich, K. P., Fedynsky, V. V. On the devastation effect of meteoritic impacts. – *Reports of the Academy of Sciences of the USSR*, New Series, 1947, Vol. 57, No. 2 (in Russian).
12. Kazantsev, A. P. The Explosion. – *Vokrug Sveta*, 1946, No. 1, p. 41 (in Russian).
13. Krinov, E. L. *The Tunguska Meteorite*. Moscow: Academy of Sciences of the USSR, 1949, p. 183 (in Russian).
14. Florensky, K. P. Some impressions about the contemporary state of the region of the Tunguska meteorite fall. – *Meteoritika*, Vol. 12, 1955 (in Russian).
15. Yavnel, A. A. On the composition of the Tunguska meteorite. – *Geokhimiya*, 1957, No. 6, p. 556 (in Russian).
16. Yavnel, A. A. op cit.; Yavnel A. A. Meteoritic substance from the place of the Tunguska meteorite fall. – *Astronomichesky Zhurnal*, 1957, Vol. 34, No. 5 (in Russian).
17. Evgeniev, I., Kuznetsova, L. The Tunguska meteorite has been found. – *Znaniye-Sila*, 1957, No. 12 (in Russian).
18. Krinov, E. L., Yavnel, A. A. The Tunguska meteorite is no longer an enigma. – *Komsomolskaya Pravda*, 1957, September 8 (in Russian).
19. See Vronsky, B. *By the Path of Kulik's: A Tale About the Tunguska Meteorite*. Moscow: Mysl, 1984, p. 34 (in Russian).
20. Florensky, K. P., Vronsky, B. I., Emelyanov, Y. M., Zotkin, I. T., Kirova, O. A. Preliminary results of the work of the Tunguska meteoritic expedition of 1958. – *Meteoritika*, Vol. 19, 1960 (in Russian), pp. 122, 126.
21. Ibid., p. 131.
22. Credit for this discovery must be given to Dr. Yury Emelyanov.
23. Florensky, K. P., Vronsky, B. I., Emelyanov, Y. M., Zotkin, I. T., Kirova, O. A. Preliminary results of the work of the Tunguska meteoritic expedition of 1958. – *Meteoritika*, Vol. 19, 1960 (in Russian), p. 120. Later, there were found at Tunguska yet more pronounced

manifestations of this effect – where the width of tree rings reached almost *two centimeters*.

24. Ibid., p. 131.
25. Ibid.
26. The leading Soviet academic organization involved in investigation of explosions, especially nuclear explosions.
27. See On the Tunguska Meteorite. – *Priroda*, 1959, No. 11 (in Russian).
28. See Staniukovich, K. P., Shalimov, V. P. On the motion of meteor bodies in the atmosphere of the Earth. – *Meteoritika*, Vol. 20, 1961 (in Russian).
29. Krinov, E. L. Where is the Tunguska meteorite? – *Priroda*, 1960, No. 5, p. 61 (in Russian).
30. See Fesenkov, V. G., Krinov, E. L. News about the Tunguska meteorite. – *Herald of the Academy of Sciences of the USSR*, 1960, Vol. 30, No. 12 (in Russian).
31. Fesenkov, V. G. On the nature of the Tunguska Meteorite. – *Meteoritika*, Vol. 20, 1961, p. 30 (in Russian).

5. Radical New Research

Science is an amazing creation of the human mind, and the science community worldwide devotes its energies to its advancement. That sounds fine, but there are problems to consider. One of the most important is the demarcation between scientific and nonscientific forms of thinking in establishing knowledge of the world about us. Of course, the aim of science is to gain scientific truth, but scientists do not have any special claim to say what is true. There exist only research standards that demand the validity of results. Not every scientific statement is a correct one, although science has developed a system of freeing itself from false but scientifically credible statements. This system is called peer reviewing. But sometimes this system works as a "voting machine" that eliminates not only ideas that are too silly for serious consideration but also those that are considered too novel for the current paradigm. Nevertheless, it is due to this system that science makes constant and reasonable progress and is not just rushing about between different and mutually inconsistent positions.

Yet this progress is being achieved under certain social and cultural conditions. Because resources allocated by society for the needs of the scientific community are limited, money will go to those scientists working on subjects within the accepted paradigm. So the lion's share goes to the "socially strong" scientists – to those with good contacts in the established institutions that allocate the money. This applies especially to the so-called "big sciences" of the twentieth and twenty-first centuries, such as particle physics and molecular biology. Big science is science needing big money to function, and this can only come from government institutions and large corporations, which are advised by the science establishment. Not unexpectedly, under such circumstances, the search for scientific truth may at times be relegated to the background.

The Soviet scientific community was *very* bureaucratized and therefore very dependent on the intellectual and moral integrity of

V. Rubtsov, *The Tunguska Mystery*, Astronomers' Universe,
DOI 10.1007/978-0-387-76574-7_5, © Springer Science+Business Media, LLC 2009

individual scientists. In that system a truly gifted scientist could pay all his or her attention to research and make really important discoveries without being distracted by the need to fight for research grants.[1] At the same time, more mediocre colleagues could also find suitable niches in the system, strangling (or at least exploiting) the gifted scholars. This feudal system of Soviet science was built under Stalin and remained practically intact until the very disintegration of the Soviet Union. The Academicians (that is, full members of the USSR's Academy of Sciences) were not just equal fellows of the scientific community. They were, first of all, the bigwigs of science, both disposing considerable amounts of money and controlling the system of rewards, such as higher degrees and prizes and the appointments of directors of research institutions. Right behind the full members of the USSR's Academy of Sciences followed in descending order corresponding members of the same Academy, full members of Academies of Union Republics, and so on. More often than not, the personal qualities of an Academician determined the whole atmosphere in the research field he or she was in charge of. If the Academician was an honest and talented person much good might be done, including the advancement of science; otherwise the harm done might be immeasurable. All the enormous achievements and no less enormous failures of Soviet science and technology were due to this system.

In theory such a system might have collapsed very swiftly, with rapacious dullards eliminating all the gifted people and occupying all the profitable positions in science. But in practice this did not happen. The number of true scientists in the Academy of Sciences of the USSR always remained considerable. This was due to two factors. First, in the 1920s and 1930s, there remained in the Academy a considerable layer of scientists who had become its members before the October coup d'état of 1917. As a whole, they maintained high intellectual and moral standards. Academician Vernadsky was an outstanding example of one of these. The second factor was the crucial role of nuclear physics in military technology after World War II. Biology did not look too important to Stalin and his mob and could be sacrificed in the name of Marxist theory. After all, Academician Trofim Lysenko solemnly promised the highest authorities of the USSR to develop a new and purely Marxist biological science that would be extremely effective and would help to breed an unbelievably high-yielding wheat. The attempt to do so appeared worth trying, although it meant liquidating classical genetics in the country. Even if the promise failed (as it did), at least representatives of the other sciences would understand who was boss. But Stalin did need the atomic bomb, which was impossible to make without real science. It couldn't be done with ideological incantations. Both Joseph Stalin and the chief of the Soviet secret police, Lavrenty Beria, who supervised the atomic project, realized this. They also understood that a dull scoundrel pretending to be a real scientist would not understand the equations of quantum mechanics and be able to use them appropriately.

Of course, freedom for the Soviet scientist in his research work was limited. While he or she was engaged in solving a problem that the State had ordered (say, developing a new thermonuclear charge) the scientist was free to pursue this search and well rewarded for success. The scientist could also put into his or her plan of scientific research work (for a five-year period, or for a year or a quarter) the themes that were of personal interest, provided this did not divert attention from the "main" task, even though rewards for successes in such fields were more modest.

However, any attempts to look into "forbidden" fields (such as conventional genetics under the reign of Academician Lysenko or problems of cybernetics in the years when it was considered in the Soviet Union as a "reactionary pseudoscience") were stopped immediately and resolutely. In the 1980s, according to official statistics, about a quarter of all scientists in the world worked in the USSR, although its population did not exceed one-twentieth of the world. Every morning, hundreds of thousands of Soviet scientific workers entered the doors of their scientific research institutes and continued to examine the recommended, or at least allowed problems. In fact, most of them were just skilled fitters at a scientific assembly line, something not foreign in other countries too.

So this "silent majority" was occupied with scientific routine, accumulating small pieces of information about the world we live in. This is necessary in itself – where else would the science geniuses find empirical data for their generalizations – but for some individuals it was not enough, and they were constantly searching for problems that would be interesting to them personally. Science had originated from simple human curiosity about the inner nature of the world around us, and only recently has it become important for production in modern society, losing simultaneously much of this early spirit of free enquiry. The Independent Tunguska Exploration Group (ITEG) that became the center of Tunguska studies for several decades arose from just this thirst for an unrestricted scientific quest. It was born half a century ago as a union of people who gathered together of their own free will, and it remains such a union.

These people proved to be gifted and purposeful. The ITEG is in some sense an exemplary scholarly community, since its members are untouched by thoughts of material or social reward for their work. On the contrary, they have spent their free time and monies earned elsewhere to satisfy their scientific curiosity concerning the enigma of the Tunguska catastrophe. What is even more important, ITEG members have been satisfying this curiosity at a highly professional level. The ITEG has become a new research organization devoted to the scientific investigation of a hard-to-solve interdisciplinary problem.

So the "inner" impetus for organizing the ITEG came from a wish for freedom of scientific investigation. But were there any other influences? Certainly, yes. It was Alexander Kazantsev's idea that the Tunguska space body (TSB) had been an extraterrestrial spaceship that prompted hundreds of professional scientists to work in this field. The launch of the first *Sputnik* in October 1957 also played an important part in this process. Just two weeks before this historic event, none other than the former British Royal astronomer Sir Harold Spencer Jones solemnly declared: "Space travel is bunk." But this first step into space made people understand that ideas that formerly looked "absurd" might in fact become a real part of life. Physicists, mathematicians, rocket engineers, and other professionals who formed the ITEG approached Kazantsev's hypothesis with rational interest. At first they simply wished to "find fragments of a spaceship," but then, when it turned out that somehow such a thing was lacking in the Tunguska taiga, their research orientation changed. Since then the ITEG has been investigating all possible traces of the Tunguska phenomenon.

A new stage in Tunguska investigations started when two teams of young Siberian scientists and engineers independently and simultaneously took an interest in the problem of the Tunguska "meteorite." These people lived and worked in the city of Tomsk, known as the "Siberian Athens." In 1878 the first university in Siberia was established in this city. By the end of the 1950s it had developed into one of the largest scientific and military-industrial centers of the region. With its population of only 250,000 there were some 25 scientific research bodies and 6 colleges, plus the State University and the Siberian Integrated Chemical Mill, which produced weapons-grade plutonium.

In October 1958 Victor Zhuravlev, a postgraduate student of Tomsk University (see Figure 5.1), visited Moscow, where he met Evgeny Krinov and Alexander Kazantsev. By that time, Krinov was already Deputy Chairman of the Committee on Meteorites of the USSR's Academy of Sciences (KMET) and Kazantsev a famous Soviet science fiction writer. Zhuravlev told them of his idea to arrange a "scientific-tourist" trip to Tunguska, and both the astronomer and the writer approved. Krinov even gave Zhuravlev a photocopy of a map of the Great Hollow and recommended that the group should try to reach the eastern border of the fallen forest area,

FIGURE 5.1. Dr. Victor Zhuravlev, a founding father of the ITEG – Independent Tunguska Exploration Group – near the epicenter of the Tunguska explosion (the ITEG expedition of 2001). Behind him one can see a "telegraph tree" – that is, a dead tree scorched and devoid of branches in 1908 as a result of the explosion, but still standing upright (*Credit*: Konstantin Shkutov, Vanavara, Russia.).

though it was still not clear where this was. The map was rather rough, but in the late 1950s it was valuable. At that time detailed maps of various regions of the Soviet Union were strictly secret.

About the same time in 1958 Gennady Plekhanov (see Figure 5.2), who worked both as a physician and as an engineer at the Betatron Laboratory of Tomsk Medical Institute, was wondering why nobody had tried to measure the radioactivity at Tunguska. If a nuclear explosion had taken place in 1908, there should still be a higher than normal level of radiation at the site of the event. Scholars, journalists, and writers had argued in newspapers and journals about this, but somehow no one had tried to check it in the field. Being experienced in measuring radioactivity, Plekhanov decided to invite some friends to go to Tunguska to settle the question once and for all. The Betatron Laboratory lacked portable radiometers, but it was rumored that the Geophysical Department of Tomsk Polytechnic Institute possessed

FIGURE 5.2. Dr. Gennady Plekhanov, the Commander of the ITEG (*Source: The Tunguska Phenomenon: 100 years of an unsolved mystery.* Krasnoyarsk: Platina, 2007, p. 44.).

such equipment, so Plekhanov visited the institute. The Geophysical Department's workers said that just a few days ago other interested people, Victor Zhuravlev and his friend Dmitry Demin, had come asking to borrow portable radiometers for a trip to Tunguska. Very soon, the two groups united to form the ITEG, which initially consisted of 12 people. So was the ITEG born. Gennady Plekhanov, then 32, became the chief of the group.

When preparing the first expedition (the ITEG-1 expedition), planned for the summer of 1959, Gennady Plekhanov, who his team called the Commander, got support from the local Party and State authorities both in Tomsk and Vanavara. Without this help it would have been difficult for them to work in the taiga. Only five years had passed since Stalin's death, and there were still concentration camps in Siberia, mainly empty but ready for any "enemies of the people." But Nikita Khrushchev's "thaw" was developing, and real people serving the monstrous state system were coming out from under its weight. When aware that the young scientists were going to the taiga to search for the remains of a hypothetical spaceship, even high-ranking Party bureaucrats began to look human and did their best to help the Tunguska researchers.

However, the expedition needed mine detectors, since in the late 1950s there still was hope that pieces of the TSB could be found with such simple instruments. (Leonid Kulik had written that some Tungus people had seen in the Great Hollow "some small pieces of silvery metal.") The military refused to give the ITEG members the detectors, which they said were secret and not available to civilians. The director of the factory producing the mine detectors told the KGB that "suspicious people" were looking for secret equipment. He wanted to know how they knew that his plant made such things? Plekhanov was summoned to Tomsk's city KGB office: "Everything that you happen to discover in the taiga will have to be immediately passed to us," they said, "especially if it is something from outer space. And a list of the expedition participants must be submitted for our approval. We forbid you to take anybody into the group without our explicit permission." Naturally, the Commander had to make a list of participants for the KGB. Strange though it may seem, that was all: the Committee for State Security neither gave permission nor prohibited the expedition to the taiga. So they left Tomsk with no official approval from the secret police. As for mine detectors, Plekhanov got some by asking the Rector of the Medical Institute Dr. I. V. Toroptsev to send an official letter to the Commander of the Siberian Military District, asking him to provide them. Some other organizations – especially Tomsk Regional Tourist Club – also helped the researchers. So the ITEG's first expedition was something like a walking tour to the taiga.

The "reconnaissance detachment," consisting of Gennady Plekhanov and Nikolay Vasilyev, left Tomsk for Vanavara on June 30, 1959, on the 51st anniversary of the Tunguska explosion. They had to inspect the route and talk with local authorities in Vanavara about aid for the expedition. (Coincidentally on the same day the American physicist Giuseppe Cocconi had sent a letter to British radio astronomer Sir Bernard Lovell, founder and director of the Jodrell Bank Experimental Station, asking him to use the world's largest steerable radio telescope to search for radio signals from extraterrestrial civilizations. Sir Bernard thought that such a search did not justify the use of the radio telescope, but that letter led directly to the start of the "SETI programs" – the search for extraterrestrial radio broadcasts, a scientific line of activity that has been extensively developed in the United States and elsewhere.)

When Plekhanov and Vasilyev arrived at Vanavara a large forest fire was raging around the settlement, and the expedition helped the native people to fight it. Ten days later, 10 other explorers joined them. In Vanavara they talked with local inhabitants, including some living eyewitnesses of the Tunguska phenomenon or their descendants. Then the expedition slowly followed Kulik's path, measuring levels of radioactivity at various points and examining the ground with mine detectors. They expected to find some fragments of a gigantic iron meteorite – or to make sure that there were no such fragments at Tunguska. By the end of July the group reached Kulik's former base at the foot of the Stoykovich Mountain. A wall calendar in a house, built some 30 years earlier, informed them that today was August 31, 1930, instead of July 31, 1959. A whole historical period had passed since Kulik visited this place.

The ITEG-1 expedition worked in the Great Hollow for 38 days, looking for abnormally high levels of radioactivity and material traces of the TSB, as well as examining the fallen forest for traces of the Tunguska forest fire of 1908 and the accelerated growth of trees. Samples of peat were taken from the swamp, and samples of wood were taken from the trees that had perished in the catastrophe and also from those that had survived. As we now know, no parts of the TSB were found. The taiga had already begun to repair and cover the consequences of the Tunguska explosion, although even half a century afterward the devastation remained discernible.

Despite the failure of the ITEG-1 expedition to find fragments from a meteorite – or a spacecraft – it did make two important discoveries. First, the level of radioactivity of soils at the center of the Great Hollow turned out to be twice that of its periphery. The level of radioactivity definitely receded in an outward direction. This was hardly a natural fluctuation. Second, in some soil samples, as well as in the ash of trees, they found an increased concentration of the rare earth elements lanthanum, cerium, ytterbium, and yttrium. Spectral analysis proved this beyond doubt. The important fact here is that rare earth elements are found in nuclear waste after atomic explosions. After the expedition, Victor Zhuravlev tried to draw the attention of specialists to this fact, but these scientists simply ignored the data. They believed that the TSB was a meteorite and therefore elements that do not occur in meteorites, such as the rare earths, cannot have anything to do with the subject. Indeed, the rare earths are not considered by astronomers as elements typical for cosmic bodies, their abundance in meteorites being about 25,000 times less than in Earth's crust. However, the explorers' research did not go unnoticed by the scholarly community and the general public of the Soviet Union and other countries. On August 28, 1959, before the expedition returned from Tunguska, the *Sovetskaya Rossiya* (*Soviet Russia*) newspaper ran an article about the unusual expedition.[2] Many other periodicals, here and abroad, soon reprinted this article. The smell of a true sensation appeared in the air. After returning from the expedition, Gennady Plekhanov found a lot of letters from interested people in the mailbox of the Betatron Laboratory. There was, for instance, a letter from Academician Vasily Fesenkov hoping to hear that the meteoritic crater had at last been found and also a letter from schoolchildren wishing to learn about the Martian space crew that had perished at Tunguska in 1908. And the well-known American newspaper the *Washington Post* asked for any unpublished materials and photographs from the taiga.

In February 1960 Gennady Plekhanov went to Moscow and Leningrad to discuss the results of the summer expedition with other scientists. First of all, he went to the KMET. Although Academician Fesenkov (the Chairman of the Committee) was absent, Evgeny Krinov and other specialists in meteoritics welcomed him. Even the increased radioactivity interested the astronomers, although they were more interested in the information about the leveled forest, the traces of the forest fire, and possible TSB substances in the soil.

Of course, it was nuclear physicists, and not meteor scientists, who could correctly evaluate the data on radioactivity. Plekhanov succeeded in meeting Academician Igor Tamm, the very scientist who had led Alexander Kazantsev to develop his idea of an extraterrestrial spaceship. Tamm was already a laureate of one Nobel and two Stalin prizes and was considered as probably the most authoritative Soviet specialist in nuclear physics. But as Plekhanov recalls, it was "just a mutually interesting talk between two colleagues."[3] Academician Tamm was fascinated with the measurements of radioactivity at Tunguska and invited the engineer to read a paper on this subject for atomic physicists.

After a short trip to Leningrad (where he got acquainted with Innokenty Suslov, the man who rescued Leonid Kulik in 1927 and whom the Suslov's crater in the Great Hollow was named after), Plekhanov went again to Moscow. There he visited the workshop in the apartment of the physicist Academician Mikhail Leontovich (1903–1981) who ran theoretical investigations in the field of controlled thermonuclear fusion at the Institute of Atomic Energy of the USSR's Academy of Sciences. Leontovich was also regarded by his science colleagues as the "Academy's conscience." For the elite of Soviet physicists, he was a model of honesty and adherence to principle. In the late 1960s and early 1970s, the highest Soviet authorities used to organize and publish in central newspapers "letters of protest" against Academician Andrey Sakharov and other dissidents in the Soviet system. It must be confessed that sometimes even worthy scientists signed such letters. But representatives of the State and Party *never* approached Mikhail Leontovich with such propositions. They knew he would call them all sorts of names. However, when in 1966 Igor Tamm and Andrey Sakharov asked Leontovich to sign a petition in defense of dissidents Yury Galanskov and Alexander Ginzburg, who were convicted on a charge falsified by the KGB, he signed it without hesitation.

Plekhanov recalls that participants of the workshop – some 20 people, mainly nuclear physicists – got together in Leontovich's apartment. They offered him their proposals, advice, and help for further Tunguska work. "Someone said: 'Mikhail Alexandrovich [Leontovich], it seems the guys are able indeed, let's give them two million rubles!' I was astounded. We in Tomsk would have been happy to get ten thousand rubles."[4] Finally, everyone agreed that a hundred thousand rubles would be an acceptable sum. Leontovich immediately phoned Academician Lev Artsimovich, the Secretary of the Branch of Physical and Mathematical Sciences of the USSR's Academy of Sciences, and the matter was settled. The Siberian branch of the academy was ordered to allocate the money for the next ITEG Tunguska expedition.

Preparations for Plekhanov's new expedition, which was to become the largest in the history of Tunguska, lasted the whole winter of 1959–1960. The main difference between the ITEG-1 and ITEG-2 expeditions was the participation of professionals. In 1960, as distinct from 1959, each research objective was assigned to a specialist. The accelerated growth of the taiga was examined by foresters from the Moscow Botanic Garden. In the Southern swamp a large team of specialists worked on the ecology of morasses. Moscow geophysicists measured levels of radioactivity under the direction of Lena Kirichenko, whose life's work was to monitor radioactive fallout after nuclear tests. Several physicists again studied the Suslov and Cranberry craters, as well as the Southern swamp and Lake Cheko (a small lake some 10 km from the epicenter of the Tunguska explosion) with new sophisticated magnetometers to see if Leonid Kulik had missed any magnetic traces. By the way, in 1999, a well-equipped Italian scientific expedition from Bologna University also studied Lake Cheko. In 1960 this attracted considerable attention among skin divers who submerged themselves in the lake, including the future cosmonaut Georgy Grechko, who subsequently participated three times in orbital flights around Earth in a *Soyuz* spacecraft and the *Salyut* orbital stations. But nothing was discovered in the lake.

The ITEG-2 expedition had 73 people working in the Great Hollow for almost two months. Small teams of two to five people would dissolve into the green sea of the taiga, but the paths of all the teams were carefully traced, and the time of their return strictly controlled. There were then no mobile phones for emergencies. The Tunguska taiga remained a savage woodland with all its dangers, including bears. But thanks to an organized system of accident prevention, there was no serious accident in the forest during all 50 years of ITEG expeditions.

Academician Sergey Korolev, a great rocket engineer and Chief Designer of Soviet spacecraft, was very interested in the work of the expedition. Under his guidance the USSR built its first intercontinental ballistic missile and the first *Sputnik*, as well as launching automatic probes to the Moon to photograph the side that is never turned toward Earth. Even when the first manned orbital spaceship *Vostok* was being tested, Korolev was thinking about future interplanetary flights, and he considered Kazantsev's hypothesis about the crash of an exploding spaceship over Tunguska in 1908 worth some attention. After all, he thought, if this hypothesis was correct and debris of the machine was found, then might some elements of its design possibly be used in terrestrial rocketry? An "enormous meteorite" seemed far less interesting to Korolev. So he actually arranged for a team of 15 scientists and technicians to search the Tunguska site for spaceship debris and also provided them with a specially equipped helicopter.

As no one found any spaceship debris, Korolev lost his interest in the Tunguska problem. The launch of *Vostok* with Yury Gagarin aboard was approaching, so there were more important issues to think about. And in the autumn of 1960 the Siberian branch of the USSR's Academy of Sciences also decided to stop supporting new expeditions to Tunguska. Why did this happen? Probably because the results of the later expeditions, which were well equipped and numerous, were no more successful than those of Leonid Kulik or Kirill Florensky. However, they did confirm that in the Great Hollow there is neither a meteorite crater nor any remains of the TSB. Magnetometric examination of the Suslov and Cranberry craters, as well as of the whole Southern swamp, convincingly demonstrated that there were no large magnetic masses present. Members of the ITEG obtained the same result after using military mine detectors to scan an enormous territory, including the hills surrounding the Southern swamp. True, there was cosmic dust in the soil, but the total mass of the spherules, calculated for the whole Great Hollow, turned out to be just about a ton – too low to be significant. Also, the presence of rare earth elements was confirmed, but in the opinion of astronomers none of these elements would likely have anything to do with the impact of a meteorite or a comet.

It also seemed that the specter of an extraterrestrial spaceship, so attractive to Academician Korolev and so repulsive to Academician Fesenkov, was again haunting Tunguska. Kirill Florensky and his colleagues from KMET, even having admitted in 1958 that the Tunguska explosion had occurred above ground, could still not believe in their own discovery and were therefore inclined to put the word "explosion" (but somehow not the word "meteorite") into quotation marks. So, according to them, what happened in 1908 over the Tunguska taiga was not an explosion in the full sense of this word. Indeed, how could a meteorite have exploded? For supporters of Kazantsev's hypothesis there was no difficulty in explaining this, but meteor specialists must have pondered the problem. A ballistic shock wave alone was not enough to explain all peculiarities of the area of the flattened forest.

There was also the question of radioactivity. Slight traces were discovered at the site in 1959, but neither Gennady Plekhanov nor his friends from the ITEG had ever dealt with measurements of radioactive fallout in the field, and they could have been wrong in their measurements. However, on the 1960 expedition professionals checked this result – scientists who had worked for years on Soviet nuclear testing grounds at Semipalatinsk and Novaya Zemlya – and they confirmed that the level of radioactivity was higher than normal, though only a little over the range of fluctuations of background radioactivity.

For the sake of the meteor specialists, Gennady Plekhanov, when writing his report on the ITEG-2 expedition for *Meteoritika* annual (*Meteoritics*, the official organ of KMET), explained the increased radioactivity at the epicenter of the Tunguska explosion as fallout from recent nuclear tests. However, repeated measurements carried out 10, 20, and 30 years later did not confirm this conclusion, since the effect has remained about the same, but with time the radioactivity after atomic and thermonuclear tests decreases considerably. Even in Hiroshima just a few years after the atomic bombing, direct measurements showed no noticeable increase of radioactivity.

Finally, the ITEG members realized that one expedition, however large and well equipped, would not solve the enigma of the

Tunguska phenomenon. There was still much work to do, and this work had to be done without direct financing from the Academy of Sciences. The lack of convincing evidence of radioactivity at the site discouraged support from the nuclear physicists, and the missing alien technology from a spaceship –nicknamed "thruster" – was also not encouraging to the Tunguska enthusiasts. But perhaps it was for the best. If the "thruster" had been discovered, the KGB would have demanded that the alien debris be handed over to them. The KGB would then have passed the "thruster" to Academician Korolev or some other significant person, and the territory of the Great Hollow would have been declared a restricted area, guarded and fenced in by barbed wire. A fantasy? Well, perhaps – as regards the thruster – but definitely not regarding the barbed wire.

Luckily enough, this problem did not arise. But ITEG leaders found themselves with another difficult task. How could Tunguska studies be advanced after the Academy of Sciences withdrew its financial backing? Actually they found a quick and effective way out of the situation: the Tunguska Exploratory Group would have to continue its research work in close collaboration with the Committee on Meteorites – the meteor specialists had not as yet lost hope of proving the normal nature of the "Tunguska meteorite," while at the same time disproving the harmful inventions of all those fantasists.

So Kirill Florensky reacted positively to Gennady Plekhanov's proposal that the next expedition to Tunguska must be organized jointly by the KMET and ITEG. The Committee on Meteorites had already allocated funds for a new expedition planned for the summer of 1961, and it now appeared an opportunity to have for the same sum a much larger number of field workers. Kirill Florensky was appointed chief of this joint KMET/ITEG Tunguska expedition. Siberian researchers responded to KMET wishes, consenting to consider closed the questions about radioactivity, for which there was dubious evidence, and the lack of pieces of meteoritic iron, which was definitely correct. Plekhanov and his friends in the ITEG group were going to map the leveled forest and examine the traces of fire as well as search for various chemical elements in the soil and water. Florensky, who followed KMET's line, planned to search for magnetic spherules but did not object to the plans of ITEG to map the leveled forest.

However, shortly before the expedition commenced, his attitude changed and he wrote to Plekhanov: "Scientific researchers in the expedition are only those who are sent in the taiga by the KMET. As for all others (that was all ITEG members), they are just auxiliary workers, bound to carry out implicitly every order of the researchers." Plekhanov replied sharply, pointing out that to solve the problem of the Tunguska phenomenon, it was the ITEG specialists who were needed – their physicists, chemists, biologists, and mathematicians. As for specialists in meteoritics who belonged to the KMET, they had no object of study in the Great Hollow because it was already evident that there was no meteorite to study. He added that if Florensky did not wish to accept complete equality for the two parts of the joint expedition – the KMET's and ITEG's – he and his colleagues would go to Tunguska on their own. Kirill Florensky retreated, and the joint expedition turned out to be a great success.

The number of participants in the joint expedition of 1961 exceeded even that of the previous expedition. There were 51 people from the ITEG and 29 from the KMET, and they remained working at Tunguska from the middle of June to early October. The main aim of this expedition was to look for cosmic dust (first of all magnetite spherules) in the soils of the region. The researchers examined territory covering 10,000 km^2 and took some 150 samples, each weighing 20 kg. At this time the expedition had no helicopter, and the samples had been collected at tens and even hundreds of kilometers from the expedition's base, so they had to be carried in rucksacks. On a bank of the Khushmo River, in the very place where 30 years previously Kulik's expedition had disembarked from its rafts (the so-called Kulik's Pier), they worked night and day to separate magnetic components from the samples before examining these under microscopes.

The chief of the expedition thought the dust produced by the Tunguska explosion must have been driven by stratospheric winds for long distances. And some dust had been discovered. It was a normal meteoritic dust, with its maximum concentration in the soil at a distance of about 80 km to the northwest from the Southern swamp. In its shape this surface structure resembled a tongue, and members of the expedition labeled it the "mother's-in-law tongue."

Alas, it was impossible to determine the date when this dust had fallen. It could go back to 1800 or to 1950. It is known that cosmic dust falls from the sky intermittently, and its concentration in different places of our planet varies considerably. If the "mother's-in-law tongue" had been reliably dated back to the year 1908, it would certainly have become a good reason for further research work in this direction. Then two questions would have been raised: what was the total mass of the meteor substance that was dispersed in this region after the Tunguska explosion and how could such a substance explode in the way it did? But with no reliable dating, it was premature for a scientist to posit these questions. First, the question that had to be answered was: when did this cosmic dust fall? A convincing answer had to be found. Alas, this did not happen.

Kirill Florensky's final report on the 1961 expedition didn't appear in *Meteoritika* until 1963.[5] During the two years that passed between the expedition and the publication of Florensky's report, he was unceasingly informing the Soviet Union – in newspapers – that the "so-called enigma" of the Tunguska meteorite no longer existed. Soon after his return to Moscow from the taiga, Florensky's articles and interviews appeared in several central newspapers in which the main result of the expeditionary works was proclaimed. It was that the substance of the TSB had been discovered – and "yes," it had been a comet. That autumn, the leading members of the ITEG sent Florensky a lot of indignant letters, refuting his claim. They pointed out that only a month previously he had personally admitted that it was premature to give any final answer to the Tunguska problem. So what had changed? Kirill Florensky maintained silence.

Despite his faults Florensky was a true scientist and a genuine Russian intellectual, and the leaders of ITEG were happy to collaborate with him, even admitting that in some of his articles he "made evident errors and even twisted facts."[6] The most likely cause of Florensky's fear of any deviation from the "Party line" in his investigations could have been the grim fate of his father, the famous Russian philosopher, theologian, and electrical engineer Pavel Florensky, who had been arrested by the NKVD ("People Commissariat of Internal Affairs" – Stalin's secret police) in 1933 and shot in 1937. Since his father was officially regarded as an "enemy of the people," Kirill had not been allowed to enter a university, and he went to the Moscow Extramural Prospecting Institute instead. Only thanks to the support of Academician Vladimir Vernadsky was he able to take up geochemical research and later to defend his doctoral thesis. Should we blame him therefore for excessive caution when his childhood and youth were spent under the Damocles sword of Stalin's system? Until 1953, members of the families of "enemies of the people" could be subjected to State repression, so Kirill Florensky had been lucky.

Of course, the ITEG's life was in some sense easier than that of KMET. The Independent Tunguska Exploratory Group, though sadly lacking academic funding, was at the same time free from any outside commanders. But any expedition sent to the taiga by the Committee on Meteorites had to report back to the Presidium of the Academy of Sciences. Therefore, the research results and conclusions must have paid due regard to the expectations of the academic chiefs. A disadvantage? Yes, for sure. But also a possible advantage. Because if these results and conclusions happened to coincide with those expectations, the researchers might hope not only for verbal approval but also for more material benefits.

At the beginning of 1962 the Moscow geophysicist and nuclear physicist Lena Kirichenko informed the Siberian Tunguska investigators that Academician Vasily Fesenkov was preparing an official conference of specialists in meteoritics at which he was planning to declare that the summer expedition of 1961 had established the final truth: the TSB had been a comet. And this outstanding discovery would be presented by the conference for a State Lenin Prize of the USSR. The prize was not overwhelming – just 10,000 rubles (about $8,000) – but the money was not the main thing. More important was the title "Laureate of the Lenin Prize of the USSR," which raised considerably the social position of its holders. The list of the main players who had found the "true explanation" for the Tunguska event must have consisted of three names: Academician Vasily Fesenkov, Kirill Florensky, and Evgeny Krinov.

Let's suppose it was a comet. Why then was the ITEG's contribution to this finding ignored? After all, it was a joint expedition, and the ITEG part was twice that of the KMET. All of them dug out, carried, separated, and studied the magnetic spherules that were considered the main proof of the cometary hypothesis. Academician Fesenkov not only refrained from inviting any of the ITEG members to the important conference but was trying to conceal from them his very intention to convene it.

There was some talk, though, that to prevent a scandal, the academician was going to include Gennady Plekhanov on the list of candidates for the State Lenin Prize.

One should, by the way, understand the procedure of presenting somebody for the State Lenin Prize in the Soviet Union. The planned conference of specialists in meteoritics was a façade, a fiction. In fact, candidates for this award were discussed and accepted by the true masters of the country, the highest Party and State bureaucrats (including some academicians) and only afterward were candidates declared as such at scientific conferences. By that time Fesenkov had obviously already enlisted the support of the authorities and all other procedures were mere technicalities.

In Tomsk, members of the ITEG regularly came together on Fridays to talk. Quite often other Tunguska researchers, from other places in the USSR – Novosibirsk, Krasnoyarsk, Moscow, Leningrad – joined them. This time, before the conference, they got together as usual. Everyone understood well that if the State Lenin Prize scheme materialized, the ITEG might forget forever about further serious studies of the Tunguska problem. The problem would have been officially solved. The devastation at Tunguska would have been caused by a comet, and that would have been the end of the matter.

In a few days Gennady Plekhanov arrived at Moscow and went to KMET, where he met his old friend Evgeny Krinov. "He was probably the most straightforward, honest, and benign person in the KMET team. The exact opposite of his immediate superior Academician Fesenkov."[7]

Krinov said: "You know that this evening Vasily Grigorievich [Fesenkov] will proclaim his plan to present the State [Lenin] Prize of the USSR to the group of scientists who solved the problem of the Tunguska meteorite by establishing its cometary nature. He is going to include Kirill Florensky and me in the list of discoverers, as representatives of the Committee on Meteorites, and yourself as the representative of the Tomsk exploratory group. So, don't you worry. Vasily Grigorievich [Fesenkov] did take into consideration your great contribution to the solution of the problem. He has already enlisted the aid of officials of the Committee on the State [Lenin] Prizes and members of the Expert Council who will certainly respond favorably."

Plekhanov replied that the problem of the TSB was far from being solved, and that there was nothing as yet worthy of any award. Evgeny Krinov was surprised. How could anyone decline the State Lenin Prize? He called upon Florensky for his aid. Perhaps the noted geochemist could persuade Plekhanov to change his mind? Florensky said that because Academician Fesenkov, the Chairman of the Committee on Meteorites, believed that the TSB was a comet, it was a comet. But somehow this did not impress Plekhanov, who promised Florensky and Krinov that if they officially attempted to nominate their cometary theory for the State Lenin Prize, then the ITEG would raise hell in the newspapers and journals. To convince the meteorite specialists that it was no joke, the ITEG published in the popular *Smena* journal a letter criticizing Florensky's position. Fesenkov and his colleagues at the Committee on Meteorites had to give up their plan to obtain the State Lenin Prize for the "Tunguska comet." But of course there could be no further collaboration between the KMET and the ITEG.

It is worth noting that Plekhanov had never been a fanatical supporter of Kazantsev's spaceship hypothesis. Having started his Tunguska investigations to verify that hypothesis, he in time came to the conclusion that the accumulated data testified against Kazantsev. In the summer of 1962, Plekhanov presented a paper at the Tenth Meteoritic Conference in which he returned to an old theory that had formerly been put forward by the Belgian astronomer Félix de Roy and Vladimir Vernadsky in Russia. According to this theory, the TSB might have been a "dense compact cloud of cosmic dust." This idea did not find much support, but Plekhanov should be praised for his integrity in turning down the State Lenin Prize and for his bravery in confronting the scientific establishment. His behavior was that of a true scientist.

In 1962, Kirill Florensky went to Tunguska without any Siberian researchers. Soon he made sure that the "mother's-in-law tongue," the shape of the distribution of space dust discovered by the joint expedition of 1961, stretched from the epicenter of the event to the northwest for a distance of more than 250 km, and he decided that everything was abundantly clear: it had been the icy core of a comet that had exploded in 1908 over the taiga. This conclusion brought to a close both his and the KMET's Tunguska investigations. Subsequently, Florensky took a job in the Institute of Space Studies of the USSR's Academy of Sciences, where he studied moon soil until his death in 1982. As for the Committee on Meteorites, the expedition of 1962 was their final attempt to work at Tunguska.

Despite the rupture between KMET and ITEG, a report by Plekhanov about investigations that had been carried out in 1961 was published in *Meteoritika* three years later and refuted claims of increased radioactivity that could be dated back to 1908 in the region. He wrote: "It is found that around the epicenter [of the Tunguska explosion] exists some increase in the level of radioactivity, which is due to the fallout of recent years. Examination of parameters of the atomic decay demonstrates convincingly that the radioactive substance was brought to this region as a result of nuclear tests in 1958. No traces of artificial radionuclides from the event of 1908 have been discovered."[8]

But this was a half-truth at best. Some traces of this kind were discovered, although Plekhanov preferred not to draw the attention of meteor specialists to this issue. In his paper he also stated that there could have been no extraterrestrial spacecraft or even a natural solid body that caused the devastation at Tunguska. Rather, it must have been a swarm of coarse particles of cosmic matter moving at a great speed. In other words, no real explosion again, just a ballistic shock wave that leveled 30 million trees in the taiga.

Since normal meteorites or even clouds of cosmic matter never seriously interested Plekhanov, the leading member of the Independent Tunguska Exploratory Group, he tried to alter the research aims of the organization. After all, he said, there are in the world so many enigmas worthy of attention and investigation. The Abominable Snowman, the lost Atlantis, the library of Ivan the Terrible concealed in the vaults of the Moscow Kremlin... Since there appear to be no fragments of an alien spaceship at Tunguska, let's find another interesting research task.

This time the ITEG said "no" to its Commander. The majority of the Exploratory Group believed that the enigma of the Tunguska meteorite had not been solved – that its dusty nature had not been verified. At the initial stage of the ITEG's existence, Gennady Plekhanov had made a considerable contribution to its formation, but now he had to leave his post – for the sake of the ITEG's future. He did resign but continued his research in the Tunguska field, and a year later discovered at Tunguska indications of genetic mutations in pines probably going back to the Tunguska explosion.

In 1963 Nikolay Vasilyev (see Figure 5.3) took the helm of the ITEG, running the group until he died in 2001, when Plekhanov took charge again. Today we can say with certainty that it was a wise choice both for the organization and for the Tunguska problem. Despite Gennady Plekhanov's later vacillations, the ITEG survived and moved on to a new stage of active life under the leadership of Nikolay Vasilyev.

FIGURE 5.3. Professor Nikolay Vasilyev (1930–2001), a member of the Russian Academy of Medical Sciences, the long-standing head of the ITEG and the leading Soviet specialist in the Tunguska problem (*Source*: Vasilyev, N. V. *The Tunguska Meteorite: A Space Phenomenon of the Summer of 1908.* Moscow: Russkaya Panorama, 2004.).

Vasilyev had been a key figure in his own field of medicine – immunology. And due to his achievements in immunological studies he was elected in 1978 a member of the Academy of Medical Sciences of the USSR (now the Russian Academy of Medical Sciences). Professor Vasilyev had run state programs on the medical and social consequences of Soviet nuclear testing at Novaya Zemlya and Semipalatinsk, as well as one dealing with the radiation problems after the Chernobyl disaster. During the 40 years that he led ITEG he transformed it from a team of enthusiastic amateurs into an informal, interdisciplinary research institute aiming at solving the enigma of the Tunguska phenomenon. Having saddled himself with the leadership of the ITEG in the early 1960s, he guided the organization both through the relatively calm periods of the 1970s and 1980s and through turbulent post-Soviet times. Even though the ITEG was a viable team, quite capable of self-organization, the energy and wisdom of Academician Vasilyev were needed to overcome many problems – large and small, external and internal – that not infrequently confronted the group.

This author was fortunate to collaborate closely with Nikolay Vasilyev in the "stormy '90s," when he moved from Tomsk to Kharkov to take a job at a large Ukrainian immunological institute. His ties with Siberian colleagues did not loosen, and soon he became the Scientific Director of the National Nature Reserve *Tungussky*, established in 1996 in Russia. But the path to the creation of this important organization protecting the Tunguska region began several decades earlier.

After their State Lenin Prize fiasco, the KMET people were inclined to rid themselves of the Tunguska meteorite affair and looked for a neutral pretext to do so. At the Tenth Meteoritic Conference (the same meeting at which Gennady Plekhanov attempted to reanimate the cosmic-dust model of the TSB), the Siberian scientists proposed to establish under the aegis of the Siberian branch of the USSR's Academy of Sciences a Commission on Meteorites and Cosmic Dust, which would have taken official responsibility for the Tunguska problem. Fesenkov, Krinov, and their colleagues understood that it would be a cover organization for the ITEG, but wisely agreed to the idea. Consequently the conference applied to the Presidium of the Academy of Sciences with an official proposal. The academic authorities knew well that a number of leading Soviet physicists were interested in the Tunguska problem and supported the nuclear hypothesis. Therefore, they immediately responded favorably to this appeal. Dr. Vladimir Sobolev, a well-known Russian geologist and an investigator of the Yakut diamond deposits, agreed to take the post of the Commission's Chairman, and Nikolay

Vasilyev and Gennady Plekhanov became his deputies and did all the work of organizing the new commission. As Vasilyev wrote several decades later, the controversy between KMET and ITEG "had been settled in a very sensible and probably the only possible way."[9] The Committee on Meteorites washed its hands of the Tunguska problem, and the Exploratory Group obtained official recognition. In 1963 the ITEG published its first collection of scientific papers, *The Problem of the Tunguska Meteorite*, which contained the findings of its expeditionary works for the preceding 5 years.[10]

Having freed themselves of the responsibility to work out the Tunguska problem any further, the meteor specialists did however reserve the right to watch over the ideological purity of this field of investigation. Of course, Kazantsev's hypothesis still remained a terrible heresy, but the ITEG, thank heavens, practically ceased to talk aloud about the "alien thruster," and on its banner were the five acceptable words for astronomers: "a cloud of cosmic dust." In his report at the Tenth Meteoritic Conference, Plekhanov even emphasized: "Our conception, explaining the Tunguska phenomenon as a collision with Earth of a cloud of cosmic dust, does not seem to differ radically from the cometary hypothesis which is being developed by Academician Fesenkov. Perhaps, there are just terminological differences which will disappear after the nature of comets is ascertained."[11]

It seemed that ITEG people were beginning to forget about their initial aspirations and that the ghost of the extraterrestrial spaceship was gradually disappearing. However, as far back as 1959 there appeared a new force in the field of Tunguska studies. This was geophysicist Alexey Vasilyevich Zolotov (see Figure 5.4), a scientific worker of the Volga-Urals branch of the All-Union Scientific Research Institute of Geophysics, who then lived and worked in the Russian town of Oktyabrsky. He did not hide the main aim of his investigations – to check up on Kazantsev's hypothesis: was the TSB an extraterrestrial spaceship that had exploded when trying to land on our planet? The first stage in his checking had to be the verification of the nuclear character of the Tunguska explosion. ITEG people generally liked Zolotov's position, but meteor specialists were utterly irritated by his investigations and bold statements. Especially shocking was the surprising fact that Zolotov's works were actively supported by the Ioffe Physical-Technical Institute (one of the largest

FIGURE 5.4. Dr. Alexey Zolotov, (1926–1995), the famous student of the Tunguska problem, who dedicated all his energy to the search for scientific proof of Kazantsev's starship hypothesis and made a very important contribution to its further development (*Source*: Plekhanov, G. F. *The Tunguska Meteorite: Memoirs and Meditations*. Tomsk: University Publishing House, 2000, p. 211.).

Soviet scientific centers of investigations in the fields of nuclear physics and nuclear chemistry), which made it possible for him to publish the results of his investigations in the *Reports of the USSR Academy of Sciences*. This journal was the most authoritative and highly rated scientific periodical in the Soviet Union. And it is well known that the place of publication of a research paper is the first and one of the most important criteria used by the scientific community to evaluate it.

As distinct from the "collectivistic" ITEG, Zolotov was an "individualist" in his studies, which both helped him (since, unlike Vasilyev, he did not need to seek a compromise among different viewpoints on the problem) and sometimes prevented him from collecting as much data as he really needed. While the ITEG was systematically gathering data about the Tunguska phenomenon, trying to build its "well-balanced" model, Alexey Zolotov was saying openly that it was, most probably, a nuclear explosion, and that an alien spaceship was not inconceivable. For KMET people this was too much, and they did their best (and worst) to discredit the scientific views that he published. Since their criticisms had little effect, they began to hurt him by methods more typical for the over-established Soviet science – in particular, by trying to stop the defense of his dissertation and publication of his scientific monograph on the Tunguska problem.

As Alexey Zolotov confessed subsequently, he had taken an interest in this problem quite accidentally. That is, some small pieces of information about the "meteorite fall" in the taiga in 1908 did reach him from time to time, but he sincerely believed that there was no special enigma in this event. But in April 1959, while working on a voluminous research report in his professional field (radiation logging of oil wells), he got so tired that he decided to seek relaxation in some easy reading. The book Zolotov came across was the recently published collection of science fiction stories by Alexander Kazantsev – *The Guest from Space* – in which was reprinted the short story of the same name that had appeared eight years earlier in the *Tekhnika-Molodyozhi* journal. Although it was not the initial source of the spaceship hypothesis (which had been published in 1946 in *Vokrug Sveta*), it proposed a tenable method for the verification of Kazantsev's idea: the searching at Tunguska for artificial radionuclides, radioactive isotopes that are formed during nuclear explosions.[12] And Zolotov suddenly had a violent urge to go to Tunguska, to take samples of soil and vegetation in the taiga, and to check these samples for radioactivity at the Volga-Urals branch of the Institute of Geophysics, where he worked, which had the necessary equipment and experienced specialists in this field. Zolotov himself worked with sources of radiation and knew well how to measure the levels of background radiation.

Alexey Zolotov was soon in action. In August 1959, when on leave, he traveled to Tunguska with his old friend Iosif Dyadkin. Of course, they had to go from the Volga to Siberia by their own means, but as geophysicists they were well paid, so they could afford the trip. Dyadkin was also an experienced specialist in nuclear geophysics (neutron and gamma-ray logging). Subsequently he became a well-known political dissident and carried out a demographical study in which he calculated how many people had perished in the gulags. His results showed that from 1928 to 1941 in the USSR 10–15 million people perished from all sorts of repression and famine. Dyadkin's paper containing these data was first distributed in *samizdat*, a system of clandestine printing and distribution of dissident literature, and then published abroad. Naturally, in April 1980 he was jailed, and those friends of his who dared to stick up for him, Zolotov included, also suffered.

But back in the summer of 1959 Zolotov and Dyadkin, having come to Vanavara, hired a small plane and made a two-hour flight over the leveled forest. In the late 1950s a flight over the taiga was no longer as difficult as it had been for Leonid Kulik, who in the late 1920s waited for years for an airplane. The flattened taiga impressed Zolotov very much, convincing him that the TSB had in fact exploded in the air. Having landed in Vanavara and rested, the friends set out by land and by August 31 reached the epicenter of the explosion. Here they explored the fallen trees for several days, collecting samples of soil and wood. Some wood samples were burned on the spot, since the radioactive substances would remain in the ash. In this way, the useful mass of the samples brought from the taiga increased considerably.

After returning to Oktyabrsky, Zolotov spent several weeks examining the ash, wood, and soil with the equipment in his institute. Simultaneously, he was writing a report about the expedition in which he described his and Dyadkin's observations of the traces of the post-catastrophic fire and the abnormally increased restoration of the forest. By the end of December 1959, Zolotov finished his measurements and completed his report, after which it was simultaneously sent to the Physical and Mathematical Branch of the USSR's Academy of Sciences and to the Committee on Meteorites.

Each responded rather differently. The KMET reviews completely rejected Zolotov's work as having no scientific value. The longest review was by Kirill Florensky, who stated, in particular, that even the fact of the overground explosion was not established beyond doubt. It appears that Florensky still could not believe his own eyes and the results he himself had obtained from the 1958 expedition.[13] As for the physicists at the Academy of Sciences, they invited Zolotov to a special conference devoted to Zolotov's investigations. This took place in January 1960 at the Physical and

Mathematical Branch of the USSR's Academy of Sciences. Zolotov read a paper in which he described his work in every detail. The conference participants adopted a resolution of complete approval of his research strategy and methods of investigation and recommended that he continue searching for artificial radionuclides in the Tunguska taiga.

However, the KMET people were not persuaded by this resolution from the USSR's Academy of Sciences, and they at once began to propagate their hypothesis of a thermal explosion of the icy comet core flying in the atmosphere at the speed of 30–40 km/s. Participants of the Ninth Meteoritic Conference that was held in Kiev in June 1960 also "blamed" Alexey Zolotov both for the way he conducted his "radioactive" research and for the results he obtained. "Zolotov's group," stated the meteor specialists, "has demonstrated an utterly irresponsible approach to collection of empirical data and its interpretation. After a short stay in the Tunguska region they presented a long report containing a number of pure inventions and proving that its authors are completely lacking elementary notions of the essence of the phenomenon under investigation..."

Zolotov, who also attended this conference, argued that judging from the lack of a discernible imprint of the ballistic shock wave on the wood, the TSB had flown at a relatively low speed – not more than a few kilometers per second. For a thermal explosion this was not fast enough. Subsequently he wrote: "However, our considerations were ignored. Criticism directed at our research work was so scathing, brutal, and unjustified, that instead of making us cease our investigations, it energized me and greatly intensified my desire to continue them."[14]

And Zolotov did in fact continue his work, not a bit embarrassed by attacks from the meteor specialists, while deriving additional inspiration from the active support of leading Soviet nuclear physicists. Since the research institute where Alexey Zolotov worked was not an academic institution, being under the USSR's Ministry of Geology, the then-president of the Academy of Sciences, Academician Mstislav Keldysh, sent an official letter to the Ministry, asking that the problem of the Tunguska meteorite be incorporated into the State plan of geological scientific research works. The Minister responded positively and Zolotov became the chief of a specialized Tunguska research group, obtaining finances from the

State budget and now having an opportunity to investigate the problem in his working hours. The Scientific Council of Leningrad *Fiztekh* – the Ioffe Physical-Technical Institute – approved Zolotov's program of work, and the document was signed by the Chief Learned Secretary of the Academy of Sciences and the Director of the Academic Institute of Applied Geophysics Evgeny Fedorov (1910–1981). This geophysicist won fame in 1937 working on the first drifting station *North Pole-1*, and during World War II he managed the USSR's Hydrometeorological Service.

Thanks to government funding, Alexey Zolotov went next year to the Great Hollow in a helicopter, wearing usual street clothes and with a briefcase in his hand. It is hard to imagine what Leonid Kulik would have said had he met somebody in the taiga dressed in such a manner!

Although some traces of the radioactive fallout from the Tunguska explosion seemed to peep out here and there, it proved difficult to establish its presence. So it was necessary to gather plenty of wood samples from the trees that had survived the Tunguska explosion, or perished, and to examine these samples using the most sensitive methods of measuring the low levels of radiation. Zolotov therefore decided to transfer his main attention from the search for artificial radionuclides to an analysis of the large area of leveled forest – something that certainly existed. Zolotov believed that all the important dynamical parameters of the TSB must have been recorded in the observed pattern of forest destruction. Consequently, as a preliminary step, the researcher had to choose between the three alternatives: had the taiga been leveled by a ballistic shock wave, by a blast wave, or by both?

Zolotov preferred to start with facts, not from hypotheses, and much less from paradigmatic ways of thinking. The "meteoritic paradigm" dictated that the TSB could only be an iron or a stony meteorite or a comet core. There was a slight chance that it could have been a carbonaceous chondrite (a class of meteorites characterized by carbon contents of up to 2 percent and more) or a "dense cloud of cosmic dust" for which there was no previous evidence. However, both Alexey Zolotov and the ITEG did not rule out these models while considering other possibilities.

Zolotov and Dyadkin first met with the Siberian researchers in the Tunguska taiga as far back as the summer of 1959. A food reserve dropped for them from a plane had sunk in a bog, and the two geophysicists found themselves in a difficult position. Perhaps, not so dangerous as that which Leonid Kulik and Oswald Guelich had been in 1927, but unlike Kulik they didn't have a horse with them which could be eaten if their situation got really bad. Luckily enough, members of the ITEG-1 expedition shared their food reserves with them, so they finished their work and returned safely to Vanavara.

In the following years, Alexey Zolotov organized 12 expeditions to the Great Hollow and gathered a lot of important information about traces of the Tunguska explosion. Usually his team arrived at the taiga in the middle of August, when ITEG people were about to return, and remained there until the first snow. So, in the field they were at least not in the way of each other, and they closely collaborated when processing the collected data. It is no mere chance that the second large collection of research papers published by ITEG in 1967 and holding a prominent place in the literature on the Tunguska problem contains, in particular, four papers authored by Alexey Zolotov.[15]

From the mid-1960s, the ITEG was also leaning in its research toward real empirical data rather than to theoretical models. The Siberian scientists were exploring mutations in pines, parameters of the area of leveled forest, and chemical anomalies in the soil, as well as questioning the many eyewitnesses to the Tunguska catastrophe who were still living. In the course of these investigations, the problem of the Tunguska explosion evolved into a multidisciplinary field of investigations with its own research community and a large set of publications. As distinct from the "meteoritic establishment" (personified in the KMET), this community was ready to consider *every* hypothesis of the TSB's origin, even the nuclear one. Nevertheless, the ITEG (as well as Zolotov's group) used in their investigations absolutely normal and strictly rigorous research methods. They performed a normal scientific investigation of a highly anomalous phenomenon. This investigation can be considered a model of serious, objective science. If we associate science with these distinctive features and not with the automatic following of paradigmatic models even when they are inconsistent with the phenomena under investigation, then we are dealing here with nothing but normal science.

So by the end of the 1960s, Zolotov decided to defend a doctoral thesis, based on the results of his investigations. Here we should say that scholarly degrees in the Soviet Union were in fact conferred on scientists by the State, not by individual universities. Of course, at first a dissertation would be considered by a Scientific Council at a university or a research institute and members of such a council would decide whether or not its author deserved to receive the degree. But the final decision was approved and the certificate issued by the Higher Certifying Commission under the Council of Ministers of the USSR.

Of course, if Zolotov had wished to obtain a degree in the field of meteoritics, he would have had no chance of success. Academician Fesenkov and other members of KMET would have barred his way. That is why his specialty was "experimental physics," and the place where the thesis was defended the Leningrad *Fiztekh*. The thesis was entitled as "Estimation of physical parameters of the Tunguska phenomenon of 1908."

Data about the radioactive fallout that had supposedly occurred after the Tunguska explosion were excluded from Zolotov's thesis. He considered it, not without reason, as too raw. But even without any evidence of hard radiation, his conclusions sounded radical:

1) The TSB was moving over the area of the leveled forest with an average speed of only 1–2 km/s – not fast enough to produce the total energy of the Tunguska explosion of many megatons of TNT.
2) The forest was leveled only by the blast wave; the ballistic shock wave did not fell any tree because it was too weak – less than 1% of the whole energy.
3) The Tunguska explosion was caused by the conversion of an inner energy of some substance to mechanical energy of the blast.

These conclusions, being hardly a direct proof of Kazantsev's hypothesis, did however argue against the meteoritic and cometary hypotheses of the TSB – against KMET's position. And they must have been defended before the Scientific Council of *Fiztekh*, consisting of very competent scientists.

Usually in Soviet science the very term "*defense* of a dissertation" was somewhat metaphorical. During the defense of Zolotov's dissertation the polemics were absolutely real and sharp, and the word "defense" had its true meaning. Academician Vasily Fesenkov sent in his utterly negative review of the work, but two other full members of the Academy of Sciences – Mikhail Leontovich and Lev Artsimovich – sent in very positive (and even enthusiastic) reviews. The leading Soviet physicists did consider the nuclear hypothesis as a plausible explanation of the Tunguska phenomenon. This was a battle that Alexey Zolotov had triumphantly won. The great majority of the members of the Scientific Council of *Fiztekh* supported conferring on him the doctoral degree, not paying too much attention to the opinion of KMET specialists. And this victory opened the way for more objective studies of the Tunguska phenomenon, not limited by the "meteoritic paradigm."

It was a great personal success for Alexey Zolotov. But he certainly owed a considerable part of this success to the ITEG and to Gennady Plekhanov in particular. If they had yielded to the KMET and allowed Academician Fesenkov and his people to officially close the Tunguska question with the help of the State Lenin Prize, hardly any scholar would have dared to support Zolotov's research. And certainly there could have been no defense of a dissertation dealing with a problem the Soviet State had decreed solved.

So research on the Tunguska mystery proved to be lucky yet again. First, Leonid Kulik did not allow it to be completely forgotten by the scientific community. Then it was Alexander Kazantsev who gave a new impetus to Tunguska studies. And now, at the third stage of these studies, the ITEG and Zolotov developed a true multidisciplinary attack on the problem. The fact that 40 years have passed since Zolotov defended his dissertation and the Tunguska problem has not been solved means that the task of doing so is much more difficult than anyone thought. After the expedition of 1961 Kirill Florensky concluded: "The work of the expedition can be summarized as having virtually completed the collection of materials which will provide descriptions of all the various forms of the physical effects produced by the Tunguska meteorite on the area of the fall."[16] This was much too hasty a conclusion. The gathering of empirical data and its examination were then in their infancy. By attempting to take the "Tunguska fortress" by storm the scientists had failed, and a long period of siege lay ahead.

With time, breaks appeared in the outer walls of the fortress, and the plan of its courtyard became partly visible to the eyes of the besiegers. The "inner citadel" of the fortress – the nature of the TSB – still remained untaken, but many things had become more understandable. And many other things *less* understandable. Somehow, the number of Tunguska enigmas started to grow again – rapidly. Which ones? We will see in the following chapters.

Notes and References

1. Derek de Solla Price, the founder of scientometrics, once added to a paper of his the following note: "This paper acknowledges no support whatsoever from any agency or foundation, but then, no time wasted, either, from preparing and submitting proposals." Sounds fine, even if somewhat sad!
2. Koginov, Y. The Mystery of the Tunguska miracle. – *Sovetskaya Rossiya*, 1959, August 28.
3. Plekhanov, G. F. *The Tunguska Meteorite: Memoirs and Meditations.* Tomsk: University Publishing House, 2000, p. 64 (in Russian).
4. Ibid., p. 71.
5. See Florensky, K. P. Preliminary Results of the 1961 Joint Tunguska Meteorite Expedition. – *Meteoritika*, Vol. 23, 1963 (in Russian). By the way, in this paper he also attempted to reanimate Yavnel's findings – which had already been explained away by KMET itself as sample contaminations from the Sikhote-Alin iron meteorite.
6. Zhuravlev V. K., Zigel F. Y. *The Tunguska Miracle: History of Investigations of the Tunguska Meteorite.* Ekaterinburg: Basko, 1998, p. 44 (in Russian).
7. Plekhanov, G. F. *The Tunguska Meteorite: Memoirs and Meditations.* Tomsk: University Publishing House, 2000, p. 160 (in Russian).
8. Plekhanov, G. F. Some results of research work of the Independent Tunguska Exploration Group. – *Meteoritika*, Vol. 24, 1964 (in Russian).
9. Vasilyev, N. V. *The Tunguska Meteorite: a Space Phenomenon of the Summer of 1908.* Moscow: Russkaya Panorama, 2004, p. 25 (in Russian).
10. *The Problem of the Tunguska Meteorite.* Tomsk: University Publishing House, 1963 (in Russian).
11. Plekhanov, G. F. Some results of research work of the Independent Tunguska Exploration Group. – *Meteoritika*, Vol. 24, 1964 (in Russian).
12. See Kazantsev, A. A Guest from Space. – *Tekhnika-Molodyozhi*, 1951, No. 3, p. 34 (in Russian).

13. By the way, nobody should accuse him, in this connection, of narrow-mindedness. Quite the contrary, Florensky understood the difficulties of the normal meteoritic explanation much better than other meteor specialists.
14. Zolotov, A. On the trail of a guest from space: Fragments from a diary. – *Smena*, 1962, Nos. 17–19 (in Russian).
15. See *The Problem of the Tunguska Meteorite*. Vol. 2. Tomsk: University Publishing House, 1967, pp. 151–153, 162–186 (in Russian). Incidentally, in one of these papers Zolotov convincingly rebutted the "dusty model" of the TSB that was still cherished by Gennady Plekhanov.
16. Florensky, K. P. Preliminary Results of the 1961 Joint Tunguska Meteorite Expedition. – *Meteoritika,* Vol. 23, 1963, p. 28 (in Russian).

6. Tracks Too Large to be Seen

The Tunguska space body (TSB) may have been enigmatic, but it did not vanish into thin air. Rather it left three big keys and several smaller ones that can help scientists to unlock the door of this mystery. The first and foremost is a "mechanical" key, namely the gigantic zone of leveled forest occupying an area of some 2,150 km^2. The second, a "thermal" key, provides two items of evidence: the burn on the trees from the light flash of the explosion, which was preserved on trees that had both perished and survived, and the consequences of the forest fire produced by the explosion.

The third key is the magnetic key. Its first component is the record of a local geomagnetic storm that started several minutes after the explosion. But we also have a distinct trace of the influence of a powerful magnetic field that has remained in the soil around the Tunguska epicenter. This is the paleomagnetic anomaly covering an area of about 1,400 km^2. Little is known about this outside the Tunguska research community. Also, at the time of the explosion of the TSB, Professor Weber in Germany recorded a strange disturbance of the geomagnetic field that could be relevant.

It is remarkable that Leonid Kulik 80 years ago was well aware of these mechanical and thermal keys and noted the importance of the "magnetic" aspect of the Tunguska phenomenon. Gigantic trees that were leveled over an enormous area and the unusual burn, covering not only branches and bark of these trees but also moss on the swamps, 20 years after the catastrophe, greatly impressed the pioneer of Tunguska studies. The theoretical speculations of scholars who had never visited the Great Hollow did not convince him. Kulik preferred to ignore their opinions, which were sometimes reasonable. Of course, attributing the leveled forest to an "unusual hurricane" and the burn of the trees to an "unusual forest fire" was absurd, but regarding the "enigmatic craters," the armchair scientists knew better than Kulik. These proved to be just thermokarst holes, as we have seen.

V. Rubtsov, *The Tunguska Mystery*, Astronomers' Universe,

DOI 10.1007/978-0-387-76574-7_6, © Springer Science+Business Media, LLC 2009

However, as an empiricist aiming at concrete results, Leonid Kulik was right. By not paying attention to the various nuances and trifles he was bending his every effort to discovering the main thing: the substance of the Tunguska object. The gigantic area of the radially leveled forest was also regarded by Kulik just as another "nuance." When Evgeny Krinov, who looked around more attentively and considered the "strange craters" with more skepticism, suggested exploring the surrounding taiga in detail, he was expelled from the expedition.

Kirill Florensky's approach to the leveled forest did not differ substantially from Kulik's. Florensky said: "Forget about the fallen trees; let's search for the substance of the meteorite. And if there are no large pieces we will look for microscopic particles." Here again, from the point of view of meteoritics, Florensky was completely right. If it was just a big stone or iron meteorite that had leveled millions of trees with its ballistic shock wave, there would be nothing incomprehensible about this. Having measured the directions of some leveled trees, the participants of the expedition of 1958 made sure that the radial character of the fallen forest was perfectly recorded, so that everyone believed no further investigations were needed.

However, later on, some "hard to explain" details began to emerge. Members of the expedition ITEG-2 felt this in 1960 when they started to explore the area of the fallen forest in a systematic way. Although the trees were lying in a radial manner, the shape of the area of leveled forest looked weird. Within this area were three zones: those of standing trees (the "telegraphnik"), mass flattening (the Tunguska explosion felled almost all trees in the territory of 500 km^2), and partially flattened trees laid in a radial direction. And it was far from being elliptical, which would have been usual for a meteoritic fall.

In 1961 the joint expedition of the ITEG and KMET had even more participants than ITEG-2, and the investigation of the leveled forest could have been continued. But Kirill Florensky, the expedition chief, thought this a "senseless waste of time and effort for obtaining quite an obvious answer." Florensky believed that even after determining exact outlines of the area of the flattened forest at Tunguska no new information would be obtained, since the TSB, according to his opinion, had been a usual meteorite. Therefore, the shape of the area of leveled forest could be only elliptical (see Figure 6.1). Reality proved to be somewhat different.

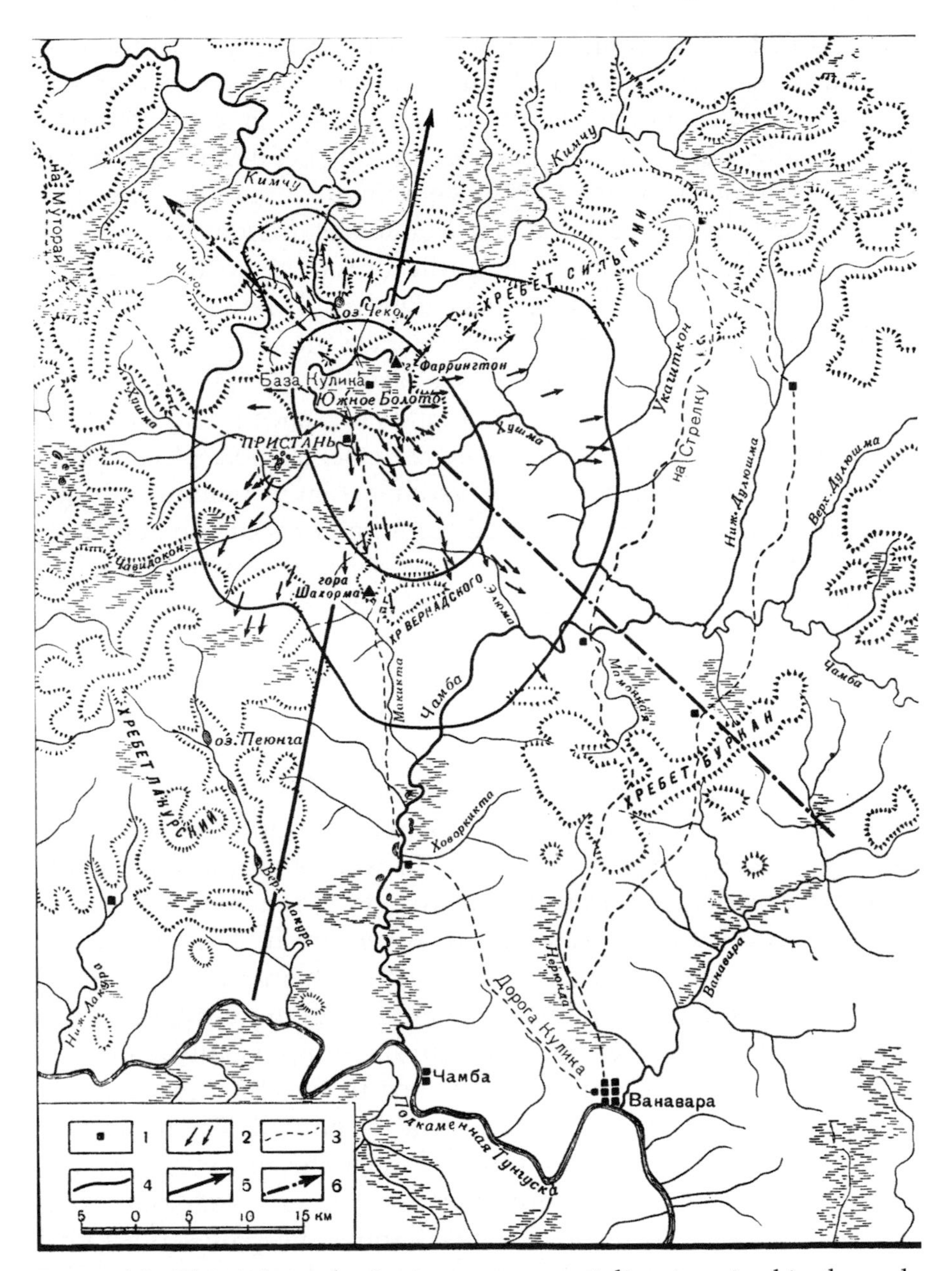

Figure 6.1. This is how the Soviet meteor specialists imagined in the early 1960s the general outlines of the area of the leveled forest (the outer closed curve) and those of the area of complete forest destruction (the inner ellipse), judging from theoretical considerations and results of the academic expedition of 1958 (*Source*: Florensky, K. P., et al. Preliminary results of the work of the Tunguska meteoritic expedition of 1958. – *Meteoritika*, Vol. 19, 1960, p. 106.).

His additional argument was, "There are tens of millions of leveled trees and to reach a reliable result it would be necessary to measure each of them. Do you think that is a sound plan?" Wilhelm Fast, a mathematician from Tomsk and an ITEG member, believed there was no need to measure the coordinates and directions of *all* the trees with precise accuracy. It would be sufficient to use small test areas, where the angles (azimuths) of the lying trees would be measured with a simple surveyor's compass accurate to 5°. It would then be possible to determine the average direction of the fallen trees very accurately. Florensky was bewildered: "Do you mean," he asked, "that if I had a hundred faulty watches I could find the exact time with the help of statistical calculations?" "Yes," Fast replied, "just so. If the number of watches is large enough and their erroneous readings are distributed according to a known statistical law, the right time may be determined with very high accuracy." Florensky yielded to this mathematical authority, although it seems he never could believe that that was so.

True, the amount of work, even limited to test areas and 5° accuracy, proved to be enormous. Needless to say, the academic Committee on Meteorites would never have been able to conduct it. The number of researchers who participated in the ITEG program "Flattened Forest" reached 120. Every summer for 20 years (from 1960 to 1979) they regularly performed their somewhat dull but highly important work. And they completed it in the nick of time – while the leveled trees were still relatively fresh. The researchers laid out more than 1,000 test areas, each of them 50 meters by 50 meters, measuring the parameters of all trees in these areas that had fallen in 1908 or perished but were still standing. Usually a test area contained from 100 to 400 or more such trees. The trees that survived the Tunguska catastrophe were also counted. The measuring treks usually lasted about two weeks through the wild sloughy taiga, with its clouds of winged bloodsucking insects – and sometimes bears. But one could not fear going astray, since the strict radial character of the leveled forest made coming back from any point to its center very easy.[1]

The northeastern sector of the leveled wood area proved to be of special interest. Previously, specialists in the Tunguska problem believed that this area did not extend in this direction farther than 4 km from the epicenter. In 1961 a team of "tree measurers," managed by Wilhelm Fast, was traveling to the northeast when they discovered to their astonishment that the forest was leveled in a northeasterly direction for up to 36 km from the epicenter. Other members of the expedition then began to help in measuring the borders. The results were traced on a map, and, step by step, before the eyes of the amazed scientists there appeared the real contour of the area devastated by the Tunguska event. Instead of an ellipse, as had been previously assumed, it resembled a gigantic spread-eagled butterfly with a "wingspan" of 70 km and a body length of 55 km (see Figure 6.2). The whole zone covered some 2,150 km^2.

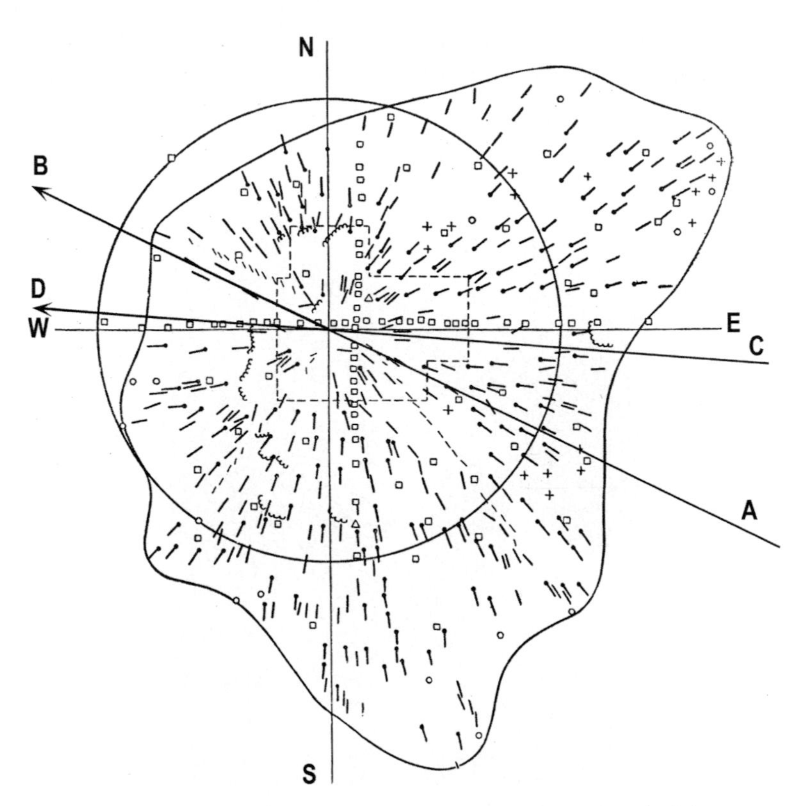

FIGURE 6.2. "Fast's butterfly": the true outlines of the leveled forest at Tunguska, 2,150 km^2 in size, according to the results of the ITEG expeditions. Lines A–B and C–D designate the first and second TSB trajectories determined by Dr. Wilhelm Fast (*Source*: Boyarkina, A. P., Demin, D. V., Zotkin, I. T., Fast, W. G. Estimation of the blast wave of the Tunguska meteorite from the forest destruction. – *Meteoritika*, Vol. 24, 1964, p. 127.).

But the ITEG members did not simply collect empirical data and plot this on maps and graphs. They immediately started to statistically process the data. It was the ITEG member Nikolay Nekrytov who first attempted to analyze the directions in which the trees had fallen, hoping thereby to determine the exact coordinates of the epicenter of the Tunguska explosion and to find a trace of the TSB ballistic shock wave.

In 1963 Wilhelm Fast (see Figure 6.3) took up this work. Fast was born in the Volga German Autonomous Soviet Socialist Republic, which existed from December 1923 to September 1941 in the USSR. In 1939 some 600,000 people lived there, two-thirds of whom were ethnic Germans, mainly descendants of those German settlers who had been invited to Russia in the eighteenth century by the Empress Catherine the Great (1729–1796). After the German invasion of the Soviet Union, the Volga German Republic was abolished and its inhabitants interned and exiled by Soviet authorities to Kazakhstan and Siberia. This is how Wilhelm Fast's family found itself in Siberia.

FIGURE 6.3. Dr. Wilhelm Fast (1936–2005), mathematician, the "Newton of Tunguska," who mapped the area of the leveled forest, preserving thereby a precise description of the most important trace of the Tunguska explosion for future generations of researchers (*Source*: Zhuravlev, V. K., Zigel, F. Y. *The Tunguska Miracle: History of Investigations of the Tunguska Meteorite*. Ekaterinburg: Basko, 1998, p. 42.).

Wilhelm had a gift for mathematics, but no real prospect of using it in exile. Luckily, after Stalin's death, he succeeded in entering the mathematical faculty of Tomsk University and, after graduating, began a doctorate course that had nothing to do with the Tunguska problem. His knowledge of the enigma was almost zero.

One day in the spring of 1960, when the ITEG people were preparing their second expedition to Tunguska, Fast accidentally attended their meeting. He listened to the enthusiasts in Tunguska studies and became interested, and subsequently helped them to translate several scholarly papers from German into Russian. After that he decided to go to the Great Hollow himself.

At first, Wilhelm was mainly engaged in measuring magnetic fields on the Southern swamp, but soon he was carried away by the imposing spectacle of the leveled forest. He even applied to his university supervisor to have his dissertation subject changed. The new subject he wanted was "Statistical parameters of the area of leveled forest at Tunguska." At first, his supervisor refused. The proposed subject seemed too far from pure mathematics. But Fast's idea was then supported by Academician Mikhail Lavrentyev (1900–1980), the first Chairman of the Siberian Branch of the USSR's Academy of Sciences. Lavrentyev was a distinguished Soviet mathematician and an outstanding specialist in the computer simulation of nuclear explosions. He had obtained a Lenin Prize for developing nuclear charges for heavy artillery, so Lavrentyev's opinion outweighed that of the supervisor and the dissertation subject was changed.

A detailed map of the leveled forest – the famous "Fast's butterfly," which was based on 650 test areas and 60,000 measured trees – was published in 1964 in KMET's *Meteoritika* annual.[2] In the following two years Wilhelm Fast successfully completed his dissertation. It was the *first* Tunguska dissertation in the world of science. Despite the misgivings of Fast's university supervisor, it turned out purely mathematical. Fast had described the statistical picture of the leveled forest most rigorously, but he believed that a mathematician should not interpret the results obtained in terms of physical models of the Tunguska event or put forward hypotheses about the TSB's nature and origin.

When the directions of the fallen trees were extended on the map toward the center of the Great Hollow they almost intersected at one point, and this looked like the epicenter of the Tunguska explosion. But Fast tried to avoid the term "epicenter." The special point, he insisted, is just a mathematical abstraction, one of the characteristics of the leveled forest area. ITEG colleagues, joking about Fast's super rigor, called this point the "epifast." It is located at a small headland on the northern bank of the Southern swamp, a couple of kilometers from the Stoykovich mountain.[3] Surprisingly, the name itself has taken root, and different variants of the Tunguska epicenter's location – proposed by various researchers – have been called "epi-" plus the first or last name of the researcher.

Fast treated the symmetrical character of the butterfly-shaped area of the leveled forest with equal caution. Its axis of symmetry ran at an angle of 115° to the east from its geographical meridian (see Figure 6.2, line A–B). It seemed quite natural to suppose that along this line – that is, from the east-southeast to the west-northwest – the TSB had been moving in the final stage of its flight. But on this subject Wilhelm Fast also preferred to refrain from any direct interpretation of his discovery. He emphasized again and again that mathematicians should not look for the physical meaning of regularities they reveal. But anyway, his calculations and conclusions could stand even the most demanding criticism.

Fast's main premise was that the trees that were affected by the Tunguska explosion could be considered as measuring instruments, whose readings are governed by certain statistical laws. And these could determine the magnitude of the force that flattened the taiga. Of course, an individual tree might not fall in a strictly radial direction, but the stronger the horizontal component of the blast wave, the smaller would be the deviation of the trees from strict radiality. Near the epicenter, the vertical component of the blast wave was predominant and therefore these deviations were considerable. Going from the epicenter to the border of the leveled forest area, we can see that its radiality becomes increasingly consistent. As we move farther from the epicenter, the vertical component of the blast wave would have become increasingly weaker, which contributes to flattening the trees in a more regular way. But farther still from the epicenter, the blast wave would have become gradually weaker so that the trees began to fall more chaotically.

Fast proved that the dynamic pressure affecting the Tunguska trees was inversely proportional to their deviations from strict radiality. So it now became possible to compose a simple formula connecting these two quantities: the force of the blast wave and the ways in which the trees had fallen. A way to the *physical* modeling of the Tunguska phenomenon was opened.

It was John Anfinogenov (Figure 6.4) who took the first step in this direction. John – a specialist in aerial photography from Tomsk – entered the ITEG in 1965 and attempted to reevaluate results obtained by Fast and to look somewhat differently at the whole picture of the Tunguska event. John's father, Fedor Anfinogenov, was, at the beginning of the 1930s, participating in the construction of the *Dneproges* – the first hydroelectric power station in the USSR. There he made friends with an American engineer and named his own son, born in 1937, after him. That is why Anfinogenov-Jr. received a name very untypical for the Soviet Union and Russia. In the ITEG Anfinogenov began to study those materials that other Tunguska specialists ignored or simply could not examine due to the lack of personnel or time. In particular, the ITEG had aerial

FIGURE 6.4. John Anfinogenov, an eminent Tunguska investigator, who has participated in 18 ITEG expeditions since 1965 and composed the map of the area of complete destruction of the taiga (*Source*: Zhuravlev, V. K., Zigel, F. Y. *The Tunguska Miracle: History of Investigations of the Tunguska Meteorite*. Ekaterinburg: Basko, 1998, p. 135.).

photographs of the Great Hollow, taken in 1949 when the environs of the Podkamennaya Tunguska River were photographed as a part of a large State program. For several years John and his colleagues studied these images and composed a map of the area of complete destruction of the Tunguska forest – 500 km^2 in size. Here almost 100% of all trees had been felled, and the shape of this area was also butterfly-like – similar in some ways to Fast's butterfly, but in other ways different (see Figure 6.5).

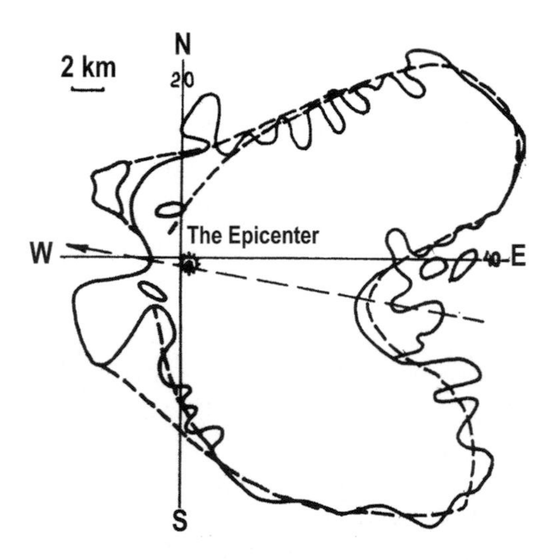

FIGURE 6.5. "Anfinogenov's butterfly" – the area of complete destruction of the Tunguska forest, 500 km^2 in size. This area has shown the most essential characteristics of the Tunguska explosion (*Source*: Zhuravlev, V. K., Zigel, F. Y. *The Tunguska Miracle: History of Investigations of the Tunguska Meteorite*. Ekaterinburg: Basko, 1998, p. 74.).

When depicting Fast's butterfly, researchers usually smoothed out its western contour, supposing that the area of the leveled forest was continuous. In fact, this supposition was wrong. There survived a strip of living trees mixed with the "telegraph poles" and running to the west directly from the epicenter. The "Anfinogenov's butterfly" does show the gap in the contour unequivocally. Its axis of symmetry does not coincide with that of "Fast's butterfly," either. True, several years later Fast himself, having studied additional data on the leveled forest collected in the field and using an improved procedure of finding the axis of symmetry, decided that the axis of symmetry of his "butterfly" must run (and, accordingly, the TSB had to fly) practically from the east to the west (see Figure 6.2, line C–D). This solution was in good accordance with the "Anfinogenov's butterfly." (Of course, it does not mean that the preceding direction of the TSB flight, determined by Fast – from the east-southeast to the west-northwest – is erroneous; rather, it may have to do with another body participating in the Tunguska event.)

The two "butterflies," which show the crucial traits of the forest leveling in the Great Hollow, are the main result from the field investigations conducted by the ITEG. Any hypotheses about the origin of the TSB and the nature of the Tunguska explosion developed without due regard for these "butterflies" would be worthless. What a pity that some scientists wishing to solve the Tunguska problem (not only European and American but Russian as well) had not the foggiest notion of these findings.

Wilhelm Fast remained active in Tunguska studies for the next 20 years, but gradually his attention shifted from science to politics and human rights. As a dissident, in 1982 Fast was expelled from Tomsk University. He met more than once with Alexander Solzhenitsyn and later became one of the founding fathers of the Tomsk branch of the *Memorial* Society.[4] But his scientific achievements cannot be overestimated. Fast's contribution toward understanding the Tunguska problem is quite comparable to that of Kulik and Kazantsev. He had fixed in figures and graphs the largest trace of the Tunguska explosion before it disappeared from the face of the Earth. And his "butterfly" is an outstanding achievement. Like Sir Isaac Newton, Fast liked to repeat: "I am not interested in hypotheses!" and he may safely be called the "Newton of Tunguska." To solve this enigma may need another Einstein, but Wilhelm Fast played his part brilliantly. He left to other specialists the task of interpreting his findings in terms of their own disciplines.

It was geophysicist Alexey Zolotov who went further. As we have seen, he attempted to interpret the structure of the area of leveled forest from a physical point of view. He reasoned that, being a material object, the TSB must have formed a ballistic shock wave, which had in its turn affected the forest before the destruction of the body itself. Somehow, Wilhelm Fast did not notice any deviations from the radial pattern of the leveled trees (neither, probably, did he try to search for them). The leveled forest area looked perfectly radial. Zolotov fully understood, however, that traces of the ballistic shock wave (the "effects of the second order") must have existed, and he set himself the target of finding these and determining from them the magnitude of the wave.

Soviet astronomer Felix Zigel (1920–1988), another contributor to Tunguska studies, illustrated the main difference between blast and ballistic shock waves, a subject of major concern to Tunguska specialists. If you throw a stone into a lake you will see how waves run from it in a concentric way. This is a good model for the blast wave produced by an explosion. Now look at a motorboat rushing across the lake. In its motion it forms a cone-like water wave that is very similar to the ballistic shock wave originating in the atmosphere from a supersonic aircraft or a meteorite.

The general scenario of the Tunguska event shared by almost all Tunguska investigators is very simple: one space body flew over central Siberia, generating in its flight a ballistic shock wave and performing no maneuvers, exploding over the Great Hollow and producing a blast wave. The TSB could, therefore, have been an ordinary meteorite, or a cometary core, or an extraterrestrial spaceship meeting disaster – any one of these would agree with this scenario. And the space body could have flown over the taiga in either a flat or a steep path, and be accompanied by either a strong or weak ballistic shock wave.

Here the term "strong" wave means that it could level trees. "Weak" means that the wave could not level them. Judging from the strict radial character of the leveled forest, we can immediately rule out the combination of a flat path with a strong ballistic shock wave. In this case, the trees would have fallen, forming a herringbone pattern and not a radial one. Therefore, only the following two physical models of the Tunguska phenomenon may be seriously considered:

1. The model with a flat TSB path, in which the magnitude of the blast wave exceeded considerably the magnitude of the ballistic shock wave.
2. The model with a steep TSB path, in which the magnitude of the ballistic shock wave is comparable to or exceeding the magnitude of the blast wave.

In both these cases the trees would have fallen radially. But how can we select from these two options by using other facts less noticeable than the radial pattern? Zolotov attempted to do that by looking at astronomical estimations of the TSB trajectory's slope. Evgeny Krinov, having studied all the evidence, came to the conclusion that it had been in the range of 5–17°. Zolotov accepted Krinov's estimation and selected the model with a flat TSB path and a weak ballistic shock wave.

Researchers believing that the forest destruction had been caused mainly by the ballistic shock wave (even if in combination with a final "thermal explosion") have preferred the model with a steep TSB path and a strong ballistic shock wave. The meteor scientists, even admitting some contribution from the explosion to the destruction of the forest, have constantly tried to minimize its magnitude. Zolotov therefore decided to calculate, from the statistical characteristics of the area of the leveled forest, the parameters of the TSB. First he attempted to find the ratio of magnitudes between the blast and ballistic shock waves. The blast wave leveled millions of trees, so, if their magnitudes were comparable, the ballistic shock wave must have leveled many of them *before* the explosion. There must therefore exist (at least where the TSB approached the Great Hollow) some fallen trees whose deviations from the radial direction are very great – up to 90°. No such deviations were found in the measured trees, however. The mean deviation was just 7.5°. From this it follows that the ballistic shock wave of the TSB *did not level even a single tree*. All trees were leveled by the blast wave only. That is, the magnitude of the ballistic shock wave was much lower than the magnitude of the blast from an explosion – less than 10% of the total energy release during the Tunguska event.

But this was just the start. Zolotov was now faced with a challenging task – to determine the exact parameters of the ballistic shock wave. The altitude of the explosion, he believed, had to be from 6 to 8 km, judging by the diameter of the zone of "telegraphnik" (standing trees). If the TSB path was flat, then its altitude of flight over the area of forest destruction was rather low, and traces of the ballistic shock wave, even if weak, could in principle be found. Although not leveling a single tree, this wave had nevertheless to alter somewhat the directions in which trees fell. That is, along the projection of the TSB trajectory, to the left and to the right of it, there must have formed a band of trees lying not strictly radially. Alexey Zolotov studied the map of the leveled forest area in detail and found two sectors where this was the case (see Figure 6.6). Here in sectors 1, 2, and 3 the blast wave was not affected by the ballistic shock wave and therefore the trees lie strictly radially. However, in sectors 4 and 5 they were deflected, when falling, by the ballistic shock wave, forming an axially symmetric structure. The axis of symmetry ran from the east-southeast to the west-northwest. The herringbone pattern was feeble, but it did exist.

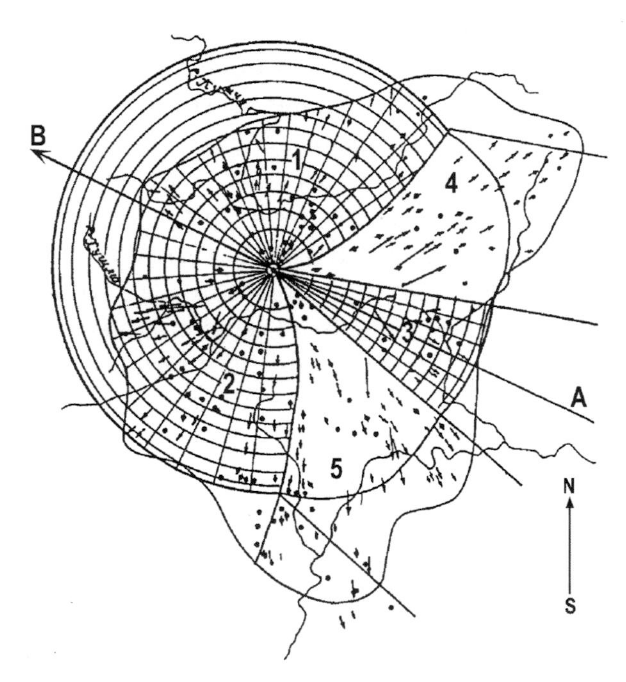

FIGURE 6.6. This shows how Dr. Alexey Zolotov determined the speed of the Tunguska space body and found the trace of its ballistic shock wave in the leveled forest. The line A–B designates the TSB trajectory according to Zolotov (*Source*: Zolotov, A. V. *The Problem of the Tunguska Catastrophe of 1908*. Minsk: Nauka i Tekhnika, 1969, p. 95.).

Because a ballistic shock wave travels symmetrically relative to the flying body's trajectory (let's remember Zigel's motorboat!), this axis is in fact the projection of the trajectory. It attests that the TSB was flying over the area of forest destruction in just this direction. This result of Zolotov's calculations concurred almost perfectly with Fast's first trajectory. What is more, having measured the deviations of the trees from radiality, Zolotov determined the real magnitude of the TSB ballistic shock wave. It approximated 7–20 kt of TNT[5] – not too little after all, but considerably less than the magnitude of the blast wave. Of course, these figures are correct only if the TSB path was flat; otherwise, the estimation has to vary.

The TSB's weak ballistic shock wave made it possible to draw strong conclusions about the dynamic characteristics of this enigmatic body – first of all about its velocity. The ballistic shock wave collided with the blast wave, forming a distinct border between the herringbone pattern and the area of the strictly radial forest leveling. Let's look again at the scheme on Figure 6.6. To find the speed of the TSB, Zolotov used the method of successive approximations. As a first approximation, he took the normal meteoritic velocity of 30 km/s. But this did not explain the location of the border between the herringbone structure and the forest, which was leveled strictly radially. After repeated calculations it was found that the velocity of the TSB was around 1 km/s, which is about the speed of the suborbital spaceplane SpaceShipOne that completed the first privately funded human spaceflight in 2004. At this velocity no "thermal explosion" – or any other type of explosion due to the kinetic energy of a moving body – is conceivable. So the TSB's explosion must have been produced by its inner energy (chemical, nuclear, or other).

It's important to note that all these values were calculated by Alexey Zolotov on the basis of strictly objective data about statistical characteristics of the "main trace" of the Tunguska phenomenon – that is, the leveled forest area. But they do depend on one important parameter of the TSB trajectory: it had to be gently sloping. The alternative model for the TSB allows that it flew "fast" and in a "steep" trajectory. John Anfinogenov decided to investigate this. He even attempted to abandon the idea of the "additional explosion" at the final point of the TSB trajectory and to explain all peculiarities of the Tunguska phenomenon in terms of "pure ballistics." Anfinogenov paid attention to the area of complete destruction of the forest, in which almost all trees had been leveled. In his opinion, this zone of just 500 km^2 contained the most reliable information about the Tunguska explosion, especially data on its magnitude, which must have been, according to his estimation, some 8 Mt of

TNT. In John's view it was not a blast. All the destruction in the taiga, he thought, must have been caused by the energy of motion of a usual (although very big) iron meteorite. Flying at an enormous speed – some 30 km/s – and naturally at a great angle to the surface – 40–50° – the meteorite formed a spindle-like ballistic shock wave, which leveled the forest strictly radially. As for the meteorite itself, it split apart, and its fragments fell down farther to the northwest, at about 5 km from the "epifast." Anfinogenov's friends immediately named this area the "epijohn."

A critic could have said that such a steep slope of the TSB trajectory does not fit the eyewitness testimonies or the well-justified figures of Krinov. But more important is that Anfinogenov's model predicted that within a relatively small zone one should find a great number of fragments of a large iron meteorite. As we know, the ITEG members have combed this zone and its environs without finding one grain of meteoritic iron.

Generally speaking, according to the scientific standards, Anfinogenov's theory should have been refuted and sent to the storehouse of many other Tunguska hypotheses – perhaps witty and sophisticated, but incapable of solving this enigma. However, John did not resign himself to defeat. Trying to explain the failure of the search in the "epijohn," he put forward an interesting idea. According to him, the Tunguska meteorite was not iron at all; instead, it consisted of a sedimentary rock that had been formed on another planet, being little different in its appearance from its terrestrial analogs. The so-called "Deer-stone," found by Anfinogenov himself on Stoykovich mountain (*not* at the epijohn), could be, in his opinion, one of these "anomalous meteorites" (see Figure 6.7). Somehow, meteor specialists do not seem as yet interested in this idea, nor hypersonics specialists in the "purely ballistic" models of the Tunguska event.

The point is that such models have been convincingly refuted – by calculations and modeling experiments. There are strong grounds for believing that the "final explosion" made a considerable contribution to the destruction of Tunguska taiga. Academician Victor Korobeynikov (1929–2003), a noted specialist in the physics of explosion, has developed with his colleagues a mathematical model and techniques to calculate the system of blast waves that are formed when large meteors fly into and explode in the

FIGURE 6.7. The enigmatic "Deer-stone," found in 1972 by John Anfinogenov on the Stoykovich Mountain, near the epicenter of the Tunguska explosion. It measures 2.5 × 1.7 × 1.2 meters and weighs more than 10 tons (*Credit*: Dr. Stanislav Kriviakov, Tomsk, Russia.).

atmosphere. In essence, they managed to deeply mathematize and advance meteoritics as a scientific field of research. During 12 years, these specialists were developing various models of the Tunguska phenomenon, testing them on fast computers and comparing the results with the real structure of the leveled forest area. It is known that the so-called inverse problems in theoretical mechanics (in which we must reconstruct the initial system of acting forces starting from the results of their action) may have more than one mathematically correct solution. For example, as we saw above in the Tunguska problem, an object flying in a flat path and producing a weak ballistic shock wave would have created more or less the same destruction in the taiga (having exploded due to its inner energy) as another object flying in a steep path and producing a strong ballistic shock wave. Academician Korobeynikov and his collaborators came to the conclusion that it was an object flying in a steep path that caused the destruction at Tunguska.

Of course a more powerful explosion occurring at a greater altitude would have produced the same effects as a less powerful one at a lower altitude. The researchers accepted that, judging from the mean diameter of the zone of standing trees, the altitude of the explosion was about 6.5 km. The results of Korobeynikov's computations are as follows: the butterfly-like shape of the leveled forest area and its radial pattern may be reproduced in calculations if the slope of the TSB trajectory was assumed to be 40°. The TSB velocity must have been 25–35 km/s and the magnitude of the *blast* wave one and a half megatons of TNT, with the magnitude of the *ballistic shock* wave three times higher. True, the calculated diameter of the zone of "telegraphnik" turned out only "about 3 km" – whereas in reality it is up to 8 km.[6] And somehow the researchers believed that the 40-degree slope was in agreement with eyewitness accounts.

In fact, both Korobeynikov's and Anfinogenov's 40° sharply contradict these accounts. Krinov's limitation of the slope of the TSB path to 17° is well justified, and this adds strength to Zolotov's model. Of course, as far as pure mathematics is concerned, Korobeynikov's calculations are sound. But astronomer Vitaly Bronshten (1918–2004), who had been studying the Tunguska problem closely for 40 years, once made an apt remark: if we are trying to unveil the real Tunguska mystery, and not just solve an abstract mathematical problem, we must reject those solutions that are inconsistent with observational data.[7]

The simplest scenario for the Tunguska event involves one body, one explosion, and no maneuvers. But strictly speaking this is just one possibility. John Anfinogenov cast doubt on its validity when he proved that the border of the leveled forest area is open in the west, although a closed line had been drawn with certitude on maps for many years. But that line had been obtained by the use of statistics, and as everyone knows there are three kinds of lies – lies, damned lies, and statistics. Individual peculiarities of a phenomenon (in our case, the area of the leveled forest) may be as important as its overall characteristics. It is no mere chance that Wilhelm Fast, when analyzing the general structure of the leveled forest area and smoothing out its contour, at first could not detect the feeble herringbone pattern in its east-southeastern part. Alexey Zolotov found it only because he knew what he was searching for and was attentive to details.

Later, it turned out that another herringbone pattern, though less distinct, existed not only in the east-southeastern part of the Tunguska territory but in the western part as well. The east-southeastern band appeared, in all probability, due to the influence of the ballistic shock wave of the TSB flying over the Tunguska taiga before its explosion. But how was a similar structure formed in the western part of the area? Let's remember that the area of the leveled forest has *two* axes of symmetry – one running from the east-southeast to the west-northwest and the second running practically from the east to the west. So, were there on that summer day of 1908 over the Great Hollow *two* space bodies, and not just one, as the simplest scenario of the Tunguska event presupposes?

Assuming that the TSB was single, we meet with a complicated problem: *what* was the ballistic shock wave reflected in the western part of the leveled forest area? According to the simplest scenario, the TSB path terminated over the Stoykovich Mountain in a powerful explosion. But what if the TSB (or a part of it) could somehow survive its fiery bath and went farther and left traces in the western part of the leveled forest area? For the explosion with a magnitude of at least 40 Mt and a maximum of 50, this assumption looks rather bold, but at least this scheme does not need another space body – which would have complicated the picture of the event too much. For example, couldn't the Tunguska meteorite (a simple iron or stony space body, or the icy core of a comet) have ricocheted from the lower atmosphere?

The "ricochet hypothesis" was originally advanced in 1929 by Ukrainian astronomer Igor Astapovich. Strictly speaking, he meant what might be called a quasi-ricochet. According to his supposition, the TSB flew through the atmosphere at an escape velocity (that is, faster than 11.2 km/s) that allows any material body to overcome Earth's gravitation. Having passed over the Great Hollow at its perigee – the minimal distance from the planet – it did not stop but traveled on into space. The air resistance only slightly distorted the TSB orbit. Astronomers have in fact observed how meteorites enter and leave Earth's atmosphere, though this usually occurs at much greater heights than it did with the TSB. So this idea was not absurd. Surprisingly, four years later Astapovich himself gave up his hypothesis – thinking it unnecessary – and returned to this idea again only in 1963.[8] He believed that there was no explosion at Tunguska; instead, the forest was leveled by the ballistic shock wave of the swiftly moving cosmic body.

Other scientists have put forward similar ideas, usually trying to explain away the lack of any meteoritic substance in the Great

Hollow. It is evident, however, that to leave the atmosphere after flying over the Southern swamp, the TSB must have moved in a *very* flat path, with its slope equal to 0° exactly, so there would have been no radial leveling of the taiga. Instead, the fallen trees would have demonstrated a very distinct herringbone pattern. The idea of a TSB ricocheting off a lower layer in the atmosphere was put forward in 1984 by Dr. Evgeny Iordanishvili. However, he did not reconcile his theory with the leveled forest area in the Great Hollow.[9] Such an analysis was subsequently performed by Gennady Plekhanov.[10] Actually, if a trace of the ballistic shock wave in the leveled forest extended beyond the epicentral zone, it means that the TSB (or a piece of it) survived the explosion and continued its motion forward. Having a sufficiently great speed, it could have flown into space, but most probably it would have fallen somewhere not far from the epicenter. To help explain this, Plekhanov recalled a local earthquake that occurred on June 30, 1908, in the Yenisey taiga at the Greater Pit River, about 460 km to the west-southwest from the explosion site, as well as unpublished reports of some eyewitnesses who saw on the same morning a bolide fly over Baykit (310 km to the west-northwest). He believes that having ricocheted, a piece of the TSB (or the TSB itself) fell in this region, producing the earthquake. However, the chance of it being found there is very low, the region being so vast.

Plekhanov's idea was expressed in "qualitative" terms, without much mathematics, and looked rather attractive. But soon, mathematical calculations revealed weak spots in his considerations. ITEG members Igor Doroshin and Evgenia Shelamova tried to find out if the ricochet effect would have been physically possible – and their results have destroyed this beautiful scheme. It turns out that changing its flight direction from the descending trajectory to an ascending one, the TSB would have endured a g loading (Earth gravitation effect plus accelerative forces) exceeding the normal Earth gravitation by 5,000 times! On the one hand, no "lower layer in the atmosphere" could be dense enough to turn the TSB so sharply. On the other hand, even if this had happened, the g loading would have immediately crushed the space body. In other words, there could have been no real ricochet over the Great Hollow.

Nonetheless, the herringbone pattern extending for 20 km in the western part of the leveled forest area remains a fact, and the simplest explanation for this fact is the survival of all or part of the TSB after the explosion. No ricochet is needed, though. Doroshin and Shelamova believe that the space body (or a swarm of its debris) traveled a distance of some 20 km after the explosion and before falling to Earth relatively close to the Great Hollow, where it might be found today.[11] However, nothing of this sort has so far been discovered in this region.

Now, when the "main trace" of the Tunguska phenomenon – namely, the butterfly-like area of 2,150 km^2 of the leveled forest – is scrutinized, the simplest Tunguska scenario (one space body – one explosion – no maneuvers) proves to be at variance with the facts.

Two axes of symmetry of this area hint at two space bodies; several local epicenters, found using aerial photography, suggest several smaller explosions; and instead of a smooth TSB flight straight to the place of its disintegration there appears a ricochet or another change in the TSB flight direction. Yes, these complications make it more difficult to produce mathematical models of the Tunguska event, whose abstract character was with good reason criticized by experienced meteor specialist Dr. Vitaly Bronshten. But to unravel this mystery without paying serious attention to these facts would not be possible.

The thermal burn of the trees, generated by the light flash, is the second most important trace of this great event. Tunguska researchers are dealing in their studies with many types of thermal injuries to the Tunguska vegetation. Some types look like the normal consequences of a forest fire, but others do not. The forest fire started by the Tunguska explosion could not be called normal, either. Kirill Florensky in the expedition of 1958 came to the conclusion that the fire "originated at the point of meteorite impact and spread in the usual manner," that is, outward.[12] To say nothing about the lack of any "point of meteorite impact" in the Great Hollow, this is simply not the case. In actual fact, as was proved subsequently, the Tunguska forest fire started simultaneously over a vast territory and did not spread beyond the boundary of the area of the leveled trees. In many places it faded soon, within 24 hours.

Strange fiery injuries to the vegetation attracted the attention of Tunguska investigators from the very beginning of their work in this region. Leonid Kulik, when breaking through the taiga to the center of the Great Hollow for the first time, was astonished by traces of a strange surface burn covering all vegetation in the region. These traces were very different from the consequences of an ordinary forest fire. That is, a forest fire did also take place here, and in the eastern and southeastern directions from the epicenter the forest did burn away, but the "surface burn" was something very different. As Kulik emphasized, the majority of leveled trees were not charred; instead, they were just singed, but traces of this singeing could be seen everywhere to a distance of 10–15 km from the center of the flattened forest area. They remained even on isolated pieces of dry land separated by water, including single trees growing among the swamps.[13] Not only trees and bushes but even marsh moss had kept these fiery marks.

It was the burn and not the subsequent forest fire that destroyed crowns and injured the bark of many trees during the Tunguska explosion. Such heat-sensitive wood species as birch, aspen, alder, and also dark conifers – pine, fir, and cedar – perished almost completely; it was mainly fire-resistant larch that had survived. Igor Doroshin correctly noticed that even in the fiercest forest fires in the taiga, fir and cedar trees never perish completely, and a considerable number of these trees survived in more humid and better shielded zones.[14] But the "fiery factor" at Tunguska acted in an unusually uniform manner. Hardly anything but a light flash could have produced such results.

Of course, Leonid Kulik did not think about any "light flash": in his time such an idea did not exist. It arose only after the first atomic explosions, when a powerful emission of light proved to be one of the most striking factors of nuclear explosions. To explain the peculiar thermal injuries of the taiga vegetation, Kulik applied his favorite hypothesis about a "fiery jet of burning-hot gases and cold bodies," which, he believed, must have struck the Great Hollow when the meteorite had split apart over it. According to his observations, the thermal factor acted downward – sometimes singeing a whole tree, sometimes influencing only its upper part. He did not scrutinize the traces of the surface burn, but at least he described these traces in sufficient detail, and his descriptions are especially valuable since they were then relatively fresh. Fortunately or unfortunately, the taiga was recovering from the consequences of the light burn much faster than from other effects of the catastrophe. To have the fallen trees rot and young growth replace them, many decades were needed; but a tree that survives a forest fire heals its injuries far sooner. The "bird's claws" (broken twigs, charred fractures) that had easily been seen in the taiga to participants of Kulik's expeditions in the late 1920s could not be found by the members of the academic team of 1958 and the ITEG-1 expedition of 1959. These "claws" were accidentally rediscovered only a year later.

Nevertheless, many years of painstaking work by Tunguska investigators made it possible to unravel the situation and to prove that there had in fact been a powerful light flash over the Southern swamp. The specialists continued to argue not about this fact but only about the magnitude of the flash. What share of the whole energy of the Tunguska explosion was emitted as visible and infrared light?

To answer this question, it was necessary, first of all, to find the lost traces of the thermal burn, which had so surprised Leonid Kulik. Of course, there was no reason to mistrust him (especially as Evgeny Krinov had also seen these strange marks). But where were they now? As it turned out, many years after the Tunguska explosion the burn traces resembled fissures filled with resin, up to half a meter in length, running along the branches. When studying living larches in 1961 that had survived the Tunguska catastrophe, two ITEG members – physicist Igor Zenkin and radio engineer Anatoly Ilyin – paid attention to the unusual damage of their branches. Through their upper parts stretched long ribbon-like cracks filled with wood resin. Judging from the number of tree rings, the cambium was damaged in 1908, after which the "wounds" began to heal, forming "resin scars." It is noteworthy that all these scars faced the center of the Great Hollow.

But finding the burn traces was just the first step in this investigation. Now the researchers had to study them in detail. This was difficult and dangerous work, perhaps the most dangerous in all Tunguska research. "Burn-hunters" selected a larch some 100–200 years old facing the center of the Great Hollow and growing in open terrain: in the middle of a swamp or at the edge of the forest. Having put on homemade foot climbers, a researcher climbed up the tree some 20 meters in height, trying to reach the top. There he examined its branches, searching for those having "resin scars." After finding such a branch, its coordinates were measured, namely, the height of its location, direction, the angle between the branch and the vertical; all data being marked on the branch itself. Then the branch was cut off and thrown down. And this process was repeated many times – at 20 meters above the ground, on the treetops of larches that swayed even in a weak wind. The selected branches were sawed up into separate pieces and examined again to eliminate any possibility of a fault.

Having finally established that it was a burn injury, the samples were sent to Tomsk, Novosibirsk, and Moscow to be investigated in well-equipped laboratories. Under a microscope, the age of the branch itself and that of the injury were verified and some additional parameters measured. In this way the Tunguska "burn-hunters" have processed more than 400 larches and collected some 1,800 samples!

Experienced specialists in forestry determined that the strange injuries were due to local heating of cambium to temperatures of 65°C or more. The results obtained are in a lengthy "Catalog of Thermal Injuries of Larch Branches."[15] This research produced some interesting results. In particular, it was found that the zone of the light burn was considerably less than the zone of the leveled forest; its length is some 18 and 12 km wide. In shape it resembles an egg, the axis of symmetry being directed almost exactly from the east to the west. Also, having discovered traces of the light burn of the trees, Igor Zenkin and Anatoly Ilyin immediately realized that this data could be used both to determine the coordinates of the source of the light flash and to estimate its energy.[16] For this purpose, they selected branches with the most distinct burn injuries. Thus the position of the source of the light flash was determined by the parallactic method (cross-bearing from different points). It was located over the southern bank of the Southern swamp, at a distance of more than 2 km southeast from the "epifast," the epicenter determined by Wilhelm Fast.[17] We can therefore conclude that the center of the explosion did not coincide with the center of the light flash. Strange indeed! But at least, these two centers lie practically along the first TSB trajectory determined by Wilhelm Fast. Dmitry Demin, a founding father of the ITEG, commenting on these facts said: "The discrepancy between the centers of the explosion and light flash may testify to their spatial disconnection."[18] That is, the center of explosion was *not* the center of the light flash! Well, it appears again that the true picture of the Tunguska phenomenon goes far beyond its simplest models...

Incidentally, Demin did not restrict his consideration to this short remark. Together with his friend Vladimir Vorobyov, he attempted to check the result obtained by Zenkin and Ilyin. After all, a tree is a living body, constantly growing and changing. A "resin scar" today may not face the same direction as it did in 1908 after the light flash. So could there be another, more precise way to find the coordinates of the flash? Vorobyov and Demin looked for the thickest tree branch to have been burnt and measured its diameter. Evidently, the higher the heat in the Great Hollow during the Tunguska explosion, the thicker the branches that would have been affected. Therefore, diameters of the thickest burned branches are good indicators of the intensity of the thermal flow in different places of the area of light burn. Gathering these figures and placing them on a map we can encircle the point from which the light had been emitted. Having processed the collected data, the researchers found that the center of the light flash had been at an altitude of 7 kilometers and 2.5 kilometers to the east from the "epifast."[19]

Now, it seems that the calculations, performed by Ilyin and Zenkin, confirm the *first* TSB trajectory determined by Fast (according to which the TSB flew to the west-northwest). At the same time, the calculations performed by Demin and Vorobyov confirm the *second* Fast trajectory (the TSB flew practically to the west)! In both cases the center of the explosion is separated from the center of the light flash by a considerable distance. Again and again, the specter of a second TSB appears on the map of the Great Hollow...

By the way, the lost and found "bird's claw" proved to be very informative. As Valery Nesvetaylo, a biologist from Tomsk, found out, all of them appeared only on those broken branches that had been dead – and therefore dry – at the moment of the catastrophe. What is more, these burns formed due to a thermal stream directed upward, not downward. It looks as if the light flash first ignited dry moss, fallen pine needles, and other flammable material covering Earth's surface in the taiga, and only after that did the fire burn the ends of dry branches that had been broken by the blast wave of the explosion. This finding made it possible to understand how the forest fire had originated simultaneously over such a large territory.

The forest fire, resulting from the powerful light flash, did not go beyond the boundary of the leveled forest area. It did not even reach its boundary. However, it covered a territory that was considerably (about five times) larger than the area of the light burn (see Figure 6.8). A very strong wind blowing immediately behind the front of the blast wave scattered the burning branches and pine needles up to a distance of some 30 km from the epicenter, but after that a "reverse" mechanism came into effect. Both the fiery

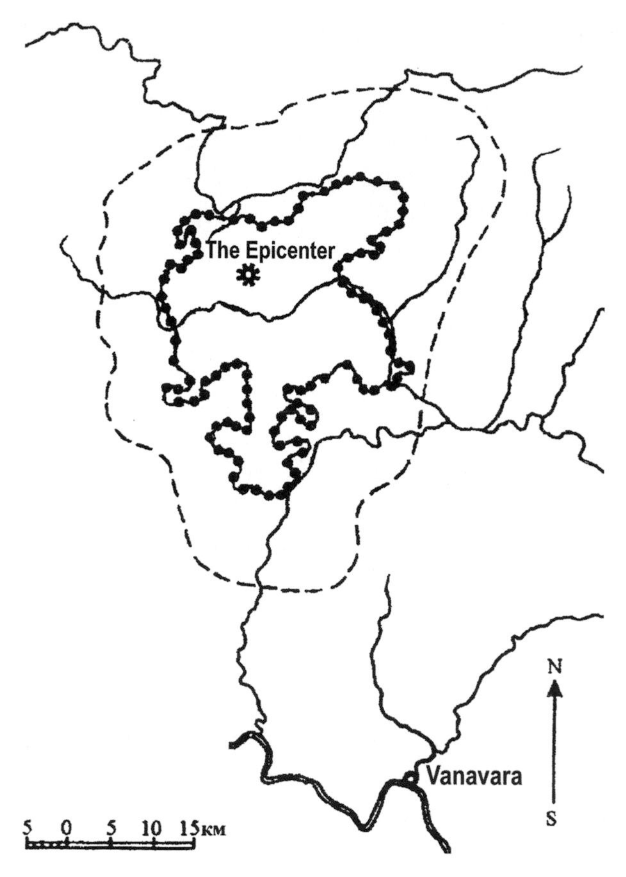

FIGURE 6.8. The zone that was occupied by the post-catastrophic Tunguska forest fire on the background of the "Fast's butterfly" (*Source*: Vasilyev, N. V. *The Tunguska Meteorite: a Space Phenomenon of the Summer of 1908.* Moscow: Russkaya Panorama, 2004, p. 137.).

ball of the Tunguska explosion and the intense forest fire near the epicenter formed a powerful pillar of hot air. The result: the strong wind changed its direction, blowing to the center of the leveled forest area. It fanned the flames of the forest fire, preventing it at the same time from spreading beyond the boundary of this area. A "fiery storm" developed, something like that which occurs when nuclear bombs are tested in the atmosphere.

As Figure 6.8 illustrates, the shape of the forest fire area is very irregular. This is understandable: the flame was spreading in this or that direction, following the terrain. Contrary to that, the burnt area from the light flash looks more regular. It may be described as

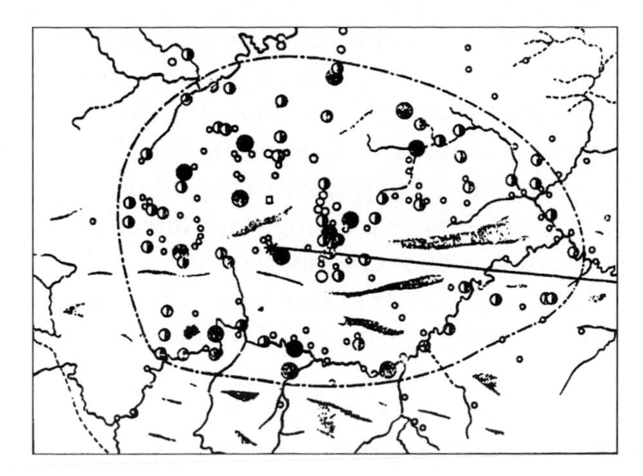

FIGURE 6.9. Smoothed outlines of the area in the Great Hollow where the vegetation was burned by the light flash of the Tunguska explosion (*Source*: Vasilyev, N. V. *The Tunguska Meteorite: A Space Phenomenon of the Summer of 1908.* Moscow: Russkaya Panorama, 2004, p. 131.).

egg-shaped, its butt-end pointing east and its pointed end toward the west (see Figure 6.9). But if we take into consideration the distribution of the intensity of the light-burn damage, a much more complicated figure arises (see Figure 6.10). It extends up to 16 km to the east from

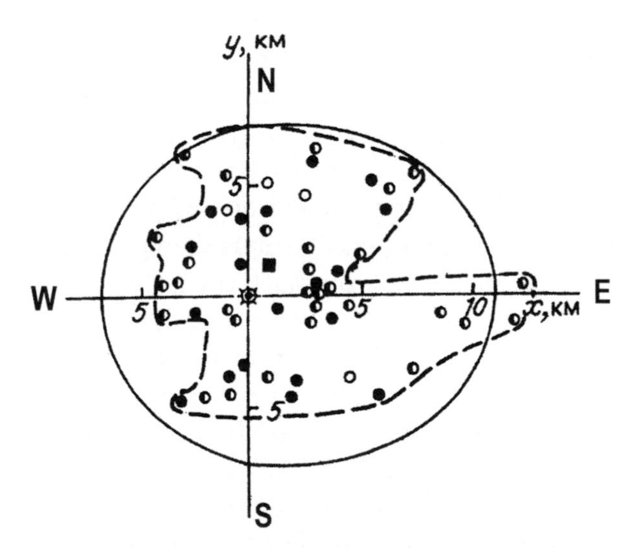

FIGURE 6.10. True (not smoothed) outlines of the Tunguska burned area from the light flash (*Source*: Zhuravlev, V. K., Zigel, F. Y. *The Tunguska Miracle: History of Investigations of the Tunguska Meteorite*. Ekaterinburg: Basko, 1998, p. 103.).

the epicenter, with two separate zones being well noticeable within it – the zone of intense burns and the zone of weak burns. Theoretically, traces of severe burning must have remained at the center of this figure and those of weak burning at its periphery. In reality the picture looks much stranger: the zone of weak burning cuts from the east into the zone of severe burning; and directly under the TSB trajectory the burning is considerably weaker than that at a distance from it. But at the very center of the figure there is evidence of the maximum level of the light flash: on a larch were found the thickest burnt branches of all.[20]

If the source of the light flash had had a regular spherical shape (as, by the way, usually happens in nuclear explosions), nothing of this sort could have taken place. Starting from the shape of the thermal burn area on the ground and using methods of computer tomography, some ITEG researchers attempted to determine the shape of the source of light emission. The result obtained by the ITEG member Stepan Razin was very peculiar: it was neither a ball, nor an egg, nor even a cylinder. The source of the light flash looked like the cap of a mushroom: a convex surface at the top and concave at the bottom.

It is worth noting that initially the idea of the light flash as the main source of the catastrophic forest fire got a hostile reception from the meteorite specialists. For them it looked too much like the "atomic heresy" – especially as the pioneer investigator of this question was the chief proponent of the nuclear hypothesis, Alexey Zolotov. However, in time all the participants in the Tunguska investigations and discussions unanimously agreed that the share of light in the total energy of the Tunguska explosion could not be less than one-tenth. But problems with the light flash were still far from being resolved. New difficulties emerged when researchers realized that the structure of the thermal burn zone was more irregular than previously thought. Near severely damaged larches, one could see other trees whose branches were quite healthy and devoid of any sign of thermal burn. In 1929, Evgeny Krinov had a similar problem when he found several groups of living trees, practically undamaged and standing not far from the epicenter. "It is incomprehensible how these small groves survived," he wrote, "since there are around them no shields against the blast wave."[21]

For the light flash this picture looked no less strange than for the blast. Mutual shielding could not explain away all cases, even taking into account that decades had passed since the catastrophe and that many traces of the thermal burn would have vanished. (Recall that Kulik saw these traces in the leveled forest area practically everywhere.) There therefore seems no escape from the conclusion that the light flash was very uneven. The intricate inner structure of the zone of thermal burn also testifies to this supposition. And last but not least, even at the epicenter of the Tunguska explosion some trees belonging to species highly sensitive to overheating – such as cedar and birch – somehow survived.

Dr. Nikolay Kurbatsky, a scientific worker of the Krasnoyarsk Institute of Forestry and a specialist in forest fires, noted that there was an evident contradiction between the severity of thermal injuries to tree branches and their final survival. To leave such scars as are still seen on Tunguska trees, the light flash must have been *very* powerful. But needles of pines, cedars, and firs die when heated to 60°C or more for several seconds. The "resin scars" testify that the Tunguska light flash was powerful enough to heat a branch one centimeter across to 65°C, at which point the cambium will die and a burn trace will appear. But in this case all the needles of the tree – and therefore the tree itself – should have perished. No living cedars, firs, and pines would have been left in the epicentral zone. In actual fact, there have remained some cedars, firs, and pines bearing no traces of the thermal burn at all. Therefore, the light emitted somehow bypassed them.[22] Two absolutely undamaged cedars grow at the western edge of the Southern swamp – practically at the epicenter. How could that happen?

Zolotov supposed that individual trees and small groves could have been shielded from the light flash by lumps of dense fog, typical in the Tunguska taiga, whose dimensions may reach tens and hundreds of meters. Hardly so. First, the undamaged trees stand, more often than not, side by side with the burnt ones. And second, the undamaged trees, as a rule, carry no noticeable structural injuries, either. The fog could probably protect the trees from the light emission – but definitely not from the blast wave. So, the "paradox of the Tunguska forest fire," formulated by Igor Doroshin, is most probably valid: a light flash with energy sufficient to ignite dry moss would inevitably have destroyed the Tunguska pines, cedars, and firs within the boundary of the light burn area. Since this is not the case, the flash must have resembled a host of powerful "thermal rays," rather than a simple fireball.

There exists, by the way, one more puzzling but little-known feature of the Tunguska forest fire that defies explanation. Leonid Kulik, emphasizing its dissimilarity from ordinary forest fires, wrote: "We do not know any other case where, after a forest fire had almost completely devastated the taiga, the dried-up trees would have been standing for 22 years, remaining so well-preserved, not darkened, but with amber-colored wood. We have been successfully using this wood as a construction material and as superb firewood."[23]

Igor Doroshin, having paid special attention to this note of the pioneer of Tunguska studies, consulted specialists in forestry and forest fires, asking them if this could have taken place? The specialists answered in unison: never! So Doroshin had to organize an excursion for them to the Great Hollow to show them the wood. Having checked that the trees did in fact perish in 1908, these specialists had to acknowledge that the Tunguska forest fire had led to the conservation of the wood and bark of the "telegraph trees." But the mechanism of this conservation still remains a mystery.[24]

Admittedly, having scrutinized the two largest traces of the Tunguska phenomenon – the areas of the leveled forest and the thermal burns – researchers did obtain a lot of valuable information, but they could not develop that information into keys to unlock the Tunguska enigma. Or rather, the keys were made but proved ineffective. They turn, so to speak, equally well in two opposite directions. Parameters of the leveled forest area correspond both to a space body of unknown nature that flew slowly in a flat path and exploded over Stoykovich mountain, and to a normal stone meteorite or to the core of a comet that flew with enormous speed in a steep path and broke apart, rather than exploded, over the same mountain. In the first case, the forest was leveled by the blast wave, in the second case by the ballistic shock wave – perhaps with a small additional blast at the very end of the TSB flight. Similarly, the powerful light flash might have been generated either by a thermonuclear explosion or by the radiance of a "super-bolide" that had been scorched hot when moving through the atmosphere. Effects of the second order (such as two axes of symmetry of the area of leveled forest or peculiarities of the zone of light burns) are certainly interesting and hint at a more intricate picture of the event, but they alone give no way of deciding between different models of the Tunguska phenomenon. So it only remains to try other locks – and other keys. Let's now turn to the magnetic key – also large, definitely important, and probably deserving more attention than was accorded to it in the past. A separate chapter will be the minimal mark of respect we can pay to this underestimated trace of the Tunguska explosion.

Notes and References

1. See Vasilyev, N. V. *The Tunguska Meteorite: A Space Phenomenon of the Summer of 1908.* Moscow: Russkaya Panorama, 2004, p. 95 (in Russian).
2. Boyarkina, A. P., Demin, D. V., Zotkin, I. T., Fast, W. G. Estimation of the blast wave of the Tunguska meteorite from the forest destruction. – *Meteoritika,* Vol. 24, 1964 (in Russian).

3. Its coordinates proved to be $60^\circ 53' 09'' \pm 6''$N and $101^\circ 53' 40'' \pm 13''$E.
4. *Memorial* is a community of several human rights organizations in post-Soviet countries – Russia, Ukraine, Kazakhstan, Latvia, and Georgia. Its main task is the awakening and preservation of the societal memory of the severe political persecution in the recent past of the Soviet Union.
5. More exactly $(6 \pm 3) \times 10^{20}$ ergs.

6. Korobeynikov, V. P., Chushkin, P. I., Shurshalov, L. V. Computing surface destruction produced by the atmospheric explosion of a meteorite. – *Cosmic Matter on the Earth.* Novosibirsk: Nauka, 1976 (in Russian).
7. See Bronshten, V. A. On some methods of calculation of the blast wave and ballistic shock wave of the Tunguska meteorite. – *Interaction of Meteoritic Matter with the Earth.* Novosibirsk: Nauka, 1980, p. 161 (in Russian).
8. Astapovich, I. S. New data on the flight of the great meteorite of June 30, 1908. – *Astronomichesky Zhurnal,* 1933, Vol. X, No. 4, pp. 465–486 (in Russian); Astapovich, I. S. The Tunguska meteorite never fell down to Earth. – *Astronomichesky Circular,* 1963, No. 238 (in Russian).
9. See Iordanishvili, E. Once again about the mystery of the "Tunguska meteorite". – *Literaturnaya Gazeta,* 1984, April 25 (in Russian).

10. See Plekhanov, G. F., Plekhanova, L. G. On a possible ricochet of the Tunguska meteorite. – *RIAP Bulletin*, 1998, Vol. 4, No. 1–2.
11. Doroshin, I. K., Shelamova, E. V. About a probable area of the fall of large debris of the Tunguska meteorite. – *The 95th Anniversary of the Tunguska Problem. Commemorative Scientific Conference. Moscow, Sternberg State Astronomical Institute, June 24–25, 2003. Abstracts of Papers*. Moscow: Moscow State University, 2003 (in Russian).
12. Florensky, K. P. Preliminary results of the 1961 joint Tunguska meteorite expedition. – *Meteoritika*, Vol. 23, 1963 (in Russian).
13. See Kulik, L. A. The leveled forest and burnt vegetation in the region of the Tunguska meteorite fall. – *Problems of Meteoritics*. Tomsk: University Publishing House, 1976, pp. 15–16 (in Russian).
14. See Doroshin, I. K. The Tunguska fiery storm. – *Tungussky Vestnik*, 2005, No. 16 (in Russian).
15. The program "Thermal Burn" was performed under the supervision of Anatoly Ilyin. Such noted scientists participated as mathematicians Boris Shkuta (Novosibirsk) and Vladimir Vorobyov (now professor and chief of the Department of Applied Mathematics of Arkhangelsk University), Evgeny Gordon (now professor and a member of the European Academy of Sciences), Vladimir Schnitke (now chief of the St. Petersburg branch of the *Memorial* Society), and many other ITEG members.
16. Zenkin, G. M., Ilyin, A. G. About the light burn of trees in the region of the Tunguska meteorite explosion. – *Meteoritika*, Vol. 24, 1964 (in Russian).
17. Geographical coordinates of the center of the light flash are as follows: 60°52′48″N, 101°55′18″E, its altitude 4,800 m.
18. Vorobyov, V. A., Demin, D. V. New results of investigation of thermal injuries of larches in the region of the Tunguska meteorite fall. – *Problems of Meteoritics*. Tomsk: University Publishing House, 1976, p. 60 (in Russian).
19. Ibid., p. 62.
20. Zhuravlev, V. K., Zigel, F. Y. op cit, p. 103.
21. Krinov, E. L. *The Tunguska Meteorite*. Moscow: Academy of Sciences of the USSR, 1949, p. 160 (in Russian).
22. See Doroshin, I. K. op cit.
23. Kulik, L. A. op cit., pp. 15–16.
24. Doroshin, I. K. op cit.

7. The Third Key

In February 1960 at the Betatron Laboratory, where the Commander of the Independent Tunguska Exploration Group (ITEG) Gennady Plekhanov worked, a thick packet arrived from Irkursk. It contained a letter from the Irkutsk Magnetographic and Meteorological Observatory, signed by the young geophysicist Kim Ivanov. This research organization had been renamed the Irkutsk Geophysical Observatory, but all the old records had been preserved in its archives. Among these materials, Ivanov had discovered a sheet of light-sensitive paper showing the disturbance of the geomagnetic field that had followed the Tunguska explosion. This was a great shock to Plekhanov and his colleagues. By that time the ITEG had been looking for about a year in vain for evidence of such an effect.

But why did the researchers believe that the Tunguska explosion had been accompanied by a magnetic disturbance? Let's look at the nature of the geomagnetic field and its interaction with the atmosphere. Although the Chinese invented the compass about 2,000 years ago, which was used by sailors and travelers for many centuries, the underlying science remained a mystery. It was the British physician and natural philosopher William Gilbert (1544–1603) who had the original thought that Earth was a gigantic magnet whose force makes the compass needle "look to the north."

Generally, magnetic fields arise around moving electrically charged particles. The magnetic field is what is called a "vector field," where not only its strength but also its direction matters. A magnetic field is measured in units called gauss and tesla, and one tesla is equal to 10,000 gauss. The strength of the geomagnetic field affecting the compass needle is only about half a gauss. So very weak magnetic fields and slight changes of their intensity are measured in nanoteslas. Geophysicists usually call one nanotesla a "gamma,"[1] so we will measure geomagnetic effects mainly in gammas.

V. Rubtsov, *The Tunguska Mystery*, Astronomers' Universe,

DOI 10.1007/978-0-387-76574-7_7, © Springer Science+Business Media, LLC 2009

The magnetic field of Earth is constantly changing, these changes being periodic and non-periodic. As a rule, compasses are not sensitive enough to feel these alterations, but magnetometers are. The non-periodic variations, which occur suddenly, are called magnetic disturbances, the most intensive and long of these being geomagnetic storms. Their amplitudes usually reach tens or hundreds of gammas, and sometimes thousands of gammas. Geomagnetic storms usually start suddenly all over the globe, lasting up to several days. These disturbances of Earth's geomagnetic field result first of all from processes occurring in the ionosphere – the upper atmosphere of our planet, which is highly ionized by the solar radiation. It begins at an altitude of about 80 km.

A geomagnetic storm is due to a surge in the speed of the solar wind, which consists of protons and electrons that constantly travel from the Sun to Earth. When penetrating the ionosphere the solar wind boosts its level of ionization, and powerful electric currents begin to flow in the upper atmosphere, producing strong magnetic fields. This leads to the total or partial fade-out of transmitted radio waves over large territories and sometimes to serious malfunctions in the work of power lines (as happened on May 13, 1980, in the Canadian province of Quebec, when 6 million people remained without commercial electric power for nine hours). There also exist the so-called substorms – occurring practically every day, sometimes globally or near globally, but too weak to affect machinery in a noticeable way.

Surprisingly, human activities can also affect the ionosphere. In 1958 American geophysicists made an unexpected discovery. It turned out that nuclear explosions could produce local geomagnetic storms in the atmosphere lasting about an hour. The separate stages of such storms lasted 10–20 min, and the intensities of the geomagnetic field reached 50 gammas. These local geomagnetic storms were first recorded in August 1958, when thermonuclear charges of some 4 Mt in magnitude were detonated over Johnston Island at altitudes of 76 and 42 km.[2] Later it was found that such effects occur only if nuclear bombs explode in the atmosphere. Even the most powerful bomb detonating at ground level leaves the geomagnetic field unchanged. Very soon, scientists uncovered the cause of this effect. It was the fiery ball of the nuclear explosion consisting of high-temperature plasma and producing hard radiation – alpha, beta, and gamma rays, as well as an increase in neutron radiation.[3] Under the influence of this radiation, the number of charged particles in the rarified air soars, and there appear in the ionosphere electric currents and magnetic disturbances.

But such plasma in the atmosphere may be formed in other ways than by nuclear explosions. In the middle of the 1940s Academician Alexey Kalashnikov discovered the magnetic effect of meteors: disturbances of the geomagnetic field accompanying the flight of meteors through the ionosphere. True, this effect lasts a few seconds at best, being much weaker than any geomagnetic storm, with amplitudes of only a fraction of one gamma.[4] Nonetheless, the nature of this phenomenon is basically the same as the nature of the nuclear geomagnetic effect.

Naturally, this brings up the question of whether a magnetic meteor effect occurred in 1908? If relatively small bolides and meteors do produce such an effect, then the enormous Tunguska space body (TSB) must have done so – in a big way. Judging from eyewitness accounts, published in Siberian newspapers, the space body approached Tunguska from the south. At a distance of about 970 km to the south-southeast from the Great Hollow lies Irkutsk and the Irkutsk Magnetographic and Meteorological Observatory, which is so important in this story, since it was separated from the TSB trajectory by a relatively short distance and could have recorded such an effect.

The idea to look for this effect occurred to Kim Ivanov in the summer of 1959. Ivanov was already aware of the artificial geomagnetic storms produced by high-altitude nuclear explosions, and he saw an opportunity to choose between the nuclear and meteoritic explanations for the Tunguska event.[5] If it were a nuclear explosion, it would have generated a geomagnetic disturbance similar to that which occurred in the Pacific in August 1958. No meteorite, however great, could produce such a local geomagnetic storm. According to the laws of physics, it could only be generated by ionizing radiation from the fiery ball of a high-altitude nuclear explosion. This fact has been established beyond doubt by American geophysicists who monitored the nuclear tests in the Pacific in 1958. But if the TSB were a huge piece of stone or iron from space, its flight would have been accompanied only by the usual magnetic meteor effect.

Luckily enough, at the Irkutsk Observatory, variations of the geomagnetic field had been recorded since 1905 on a 24-hour basis. So on the morning of June 30, 1908, the magnetometers did record a noticeable disturbance of the geomagnetic field. And this disturbance differed radically from a meteor magnetic effect. It started *after* the Tunguska explosion and lasted about five hours. Let's remember that a magnetic meteor effect occurs *during* a meteor's flight and lasts just several seconds. So, Kim Ivanov had discovered just that geomagnetic effect which had been recorded at the Irkutsk Magnetographic and Meteorological Observatory, but either missed or ignored by Dr. Arkady Voznesensky, the then Director of the Observatory. And what is no less unusual, there was on the magnetograms no sign of the "normal" magnetic meteor effect. For such a gigantic bolide this is very strange, and we can therefore suppose that the TSB flew at a low velocity not only over the Great Hollow but also through the ionosphere, its speed not being sufficient to have a vast plasma envelope form around it.

So, there was no disturbance of the geomagnetic field usually accompanying the flight of meteors. But what was there instead? The Irkutsk magnetogram is reproduced in Figure 7.1. During seven hours before the explosion of the TSB, the geomagnetic field was very

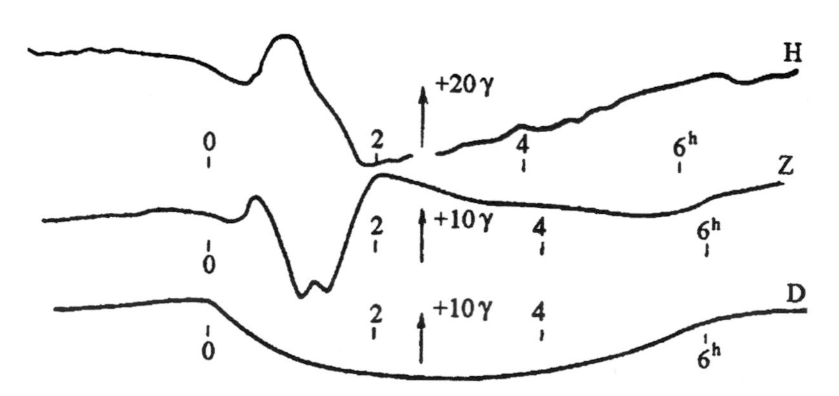

FIGURE 7.1. The geomagnetic storm, dated June 30, 1908, as recorded by instruments of the Magnetographic and Meteorological Observatory at Irkutsk. It started several minutes after the unknown space body exploded over central Siberia and was similar to the geomagnetic disturbances following nuclear explosions in the atmosphere (*Source*: Zhuravlev, V. K., Zigel, F. Y. *The Tunguska Miracle: History of Investigations of the Tunguska Meteorite*. Ekaterinburg: Basko, 1998, p. 82.).

calm. At 0 h 20 min GMT, that is, 6 min after this body exploded, the intensity of the geomagnetic field abruptly increased by several gammas and remained at that level for about 2 min. This was the initial phase of the local geomagnetic storm (or the so-called "first entry"). Then started a second phase – "the phase of rise." The geomagnetic field reached its maximum intensity at 0 h 40 min GMT and remained at the same level for the next 14 min. It then began to drop, the amplitude decreasing for some 70 gammas. It returned to its initial undisturbed level only five hours later.[6]

These four stages, the first entry, the phase of rise, the phase of fall, and the phase of relaxation, are also typical of usual solar magnetic storms. However, during a solar geomagnetic storm the first entry lasts 30 minutes on average, whereas in Irkutsk it lasted two minutes only. The third (main) phase of the solar magnetic storm usually lasts 5 to 10 hours. On June 30, 1908, this phase was much too short – just one and a half hours. And finally, the relaxation phase usually lasts 10 to 50 hours, while in the recording at Irkutsk it lasted not more than four hours. Such effects have *never* been observed by astronomers studying meteor phenomena.[7] The only parallel for this was the artificial geomagnetic storms that occurred in 1958 over Johnston Island during the high-altitude nuclear tests.

Many years later, in 1986, when talking with the ITEG member Victor Zhuravlev, Kim Ivanov confessed that he had recognized the similarity between the Tunguska geomagnetic effect and the nuclear-generated one, as well as its far-reaching implications, and had discussed this question with the author of the "spaceship hypothesis" Alexander Kazantsev and astronomer Felix Zigel. They attempted to convince Kim Ivanov that he should make this public. Kazantsev and Zigel believed that the scientific community would listen to the expert opinion of such a distinguished specialist. Yet Ivanov did not dare to do so, since he was sure that strong evidence in favor of Kazantsev's hypothesis would not only not have been accepted by established science but would have provoked plenty of protests, which would have hampered the Tunguska studies.[8]

Kim Ivanov was a serious researcher and became one of the leading Russian geophysicists. He examined the magnetograms from the Irkutsk Observatory and those from nuclear testing and wrote a paper for the Russian academic *Astronomical Journal.*

Ivanov did not offer any hypothesis for the origin of the effect discovered. He gave instrumental data and explained it – that was all. The *Astronomical Journal* rejected Ivanov's work, but the Committee on Meteorites (KMET) accepted his paper for publication in KMET's annual *Meteoritika*. They reasoned that the nature of the Irkutsk geomagnetic effect was probably vague, yet it was not a fancy finding and therefore should be published.[9]

Simultaneously with Kim Ivanov, and independently of him, the search for the Tunguska geomagnetic storm was also being pursued by the ITEG members. At that time, they were still trying to find the hypothetical "spaceship thruster." So, having heard about the nuclear geomagnetic storms, they began looking for information about the state of Earth's magnetic field during and after the Tunguska event. In 1959 Gennady Plekhanov and Nikolay Vasilyev sent inquiries to practically all geophysical observatories that had been functioning in 1908, and they received answers from 18 observatories and magnetometric stations.

For a long time these answers were disappointing: on June 30, 1908, the measuring instruments of the observatories had not recorded any disturbances. On that day the magnetic field of our planet had remained calm everywhere outside the Tunguska region. However, the magnetograms that Kim Ivanov sent to the ITEG were a true godsend, because they immediately led to a very detailed examination of those records, especially that by geophysicist Alexander Kovalevsky who had been specially invited to the ITEG to analyze the materials that were then arriving at Tomsk from Russian and foreign geophysical observatories.

Having compared the Irkutsk magnetogram with those recorded by American geophysicists during the high-altitude nuclear tests in 1958, Kovalevsky concluded that the Tunguska geomagnetic effect did not differ in any essential way from the artificial nuclear geomagnetic storms. Kim Ivanov had arrived at the same conclusion, but did not say this in his publications. It was already known that on June 30, 1908, no other magnetometric station on this planet had detected any disturbances. Therefore, the geomagnetic effect recorded at the Irkutsk Magnetographic Observatory had to be a very local effect. This was an important piece of information. Without it, one could have supposed that it had been just a simple, even if unusually short, solar geomagnetic storm. In

February of 1960, a paper entitled "On the Geomagnetic Effect of the Tunguska Meteorite Explosion" appeared in the journal *Fizika* (*Physics*) that was being issued by Tomsk University.[10] Referring to Ivanov's findings, its authors – Gennady Plekhanov, Alexander Kovalevsky, Victor Zhuravlev, and Nikolay Vasilyev – boldly likened this geomagnetic disturbance to the "artificial magnetic storms" that had followed thermonuclear explosions over the Pacific islands in 1958. Their sensational conclusion was that the "geomagnetic signatures" of the storms from both nuclear explosions and the Tunguska event were practically indistinguishable. In fact, if the only thing known about the Tunguska explosion had been its geomagnetic signature and no other traces or instrumental records had survived, we would have had to conclude that it was a nuclear explosion.

Subsequently, Kovalevsky made a great contribution to Tunguska studies, trying to find out the origin of the geomagnetic effect, looking for materials from the TSB in the soil, investigating traces of the light burn of vegetation and processing eyewitness reports. In 1979, his active research work was however interrupted for almost two years when he was flung into prison for keeping at home some dissident literature. But of course, it was not Kovalevsky who discovered the Tunguska geomagnetic effect. The true discoverer was Kim Ivanov.

It is worth repeating that not a single magnetometric station that existed in 1908 in Russia or elsewhere detected any noticeable variations of the geomagnetic field. But if it were just an unusually short solar magnetic storm that coincided by chance with the Tunguska event it would have been recorded outside Irkutsk as well. Therefore, this effect could only have been due to the Tunguska explosion. So did it mean that the Tunguska explosion could have been nuclear?

Although the ITEG researchers were looking for a geomagnetic trace of this explosion, starting from the association with similar nuclear-produced effects, it seems that Kim Ivanov's discovery had somewhat embarrassed them. Yes, they acknowledged a close similarity between the Tunguska magnetic storm and artificial magnetic storms of 1958, but they were in no hurry to declare it the final proof of the nuclear nature of the Tunguska explosion. Instead, they started to search for other, nonnuclear, explanations. This was the proper scientific approach to this question. Before accepting the nuclear explanation it had to be tested. As the famous philosopher Sir Karl Popper (1902–1995) used to say, every genuine test of a theory is an attempt to refute it. So it was necessary to look for another plausible explanation of the Tunguska magnetic storm. What else could have produced it? Could there be anything common in the thermonuclear explosions of 1958 and the Tunguska explosion of 1908, apart from possible radiation?

Certainly yes! There were shock waves! Let's remember that the magnitudes of these explosions were more or less comparable: some 4 Mt in 1958 and 40–50 Mt in 1908. Could it be the shock wave that had produced the geomagnetic effect in both cases? Independently, Kovalevsky and Ivanov developed the same hypothesis that the regional magnetic disturbance had started when the shock wave of the Tunguska explosion had struck the ionosphere.

True, even the "shock wave explanation" of the Tunguska geomagnetic effect looked from the meteoritic standpoint rather heretical, since it meant that there had occurred an *explosion* during the Tunguska event, whereas the meteorite specialists believed it had been a ballistic shock wave that had leveled the trees in the taiga. But no ballistic shock wave could have produced such a geomagnetic effect that had been recorded by the magnetometers of the Irkutsk Observatory. "Assuming that the recorded variations of the geomagnetic field were due to the ballistic shock wave of a swiftly flying meteorite," wrote Alexander Kovalevsky, "it would be impossible to explain the complicated character of these variations [of the geomagnetic field] and the time lag between the moment of the meteorite fall and the beginning of the [geomagnetic] effect."[11]

Generally, models proposed by various researchers to explain the Irkutsk geomagnetic storm are

1. Those assuming that the ionosphere was affected by the substance of the Tunguska comet's tail or by the high-temperature fiery ball that formed when its core exploded;
2. Models in which the main factor was the blast wave of the Tunguska explosion;
3. Those admitting that the geomagnetic effect was produced by hard radiation from this explosion – that is, highly penetrating alpha, beta, and gamma rays, as well as neutron radiation.

In particular, astronomers Grigory Idlis and Z. V. Karyagina accepted that the TSB "had definitely been a comet." They believed that the solar wind and comet tails are very similar. Consequently, Idlis and Karyagina supposed that the ionized comet's tail had to affect the magnetic field of Earth as does the solar wind. And since it is this wind that generates usual geomagnetic storms, the comet tail would produce a similar effect.[12] In fact, comet tails are composed of very rarified ionized gases and dust, whereas the solar wind consists of fast streams of electrons and protons. Therefore, the "Tunguska comet" tail could not produce a geomagnetic storm. Besides, from their theory it directly followed that the "cometary" geomagnetic storm would inevitably have encompassed the whole globe, as comet tails are much larger than our planet, while the localness of the Tunguska geomagnetic effect had been established beyond doubt. This is why the theory of Idlis and Karyagina failed to explain the event. Other astronomers had immediately noticed their mistake. For instance, Academician Vasily Fesenkov, even being the leading supporter of the cometary hypothesis, was not tempted by the spurious analogy between comet tail and solar wind and preferred to simply ignore the Tunguska geomagnetic effect.

Geophysicist Saken Obashev, realizing that the blast wave or comet's tail could not explain all features of the geomagnetic effect (nor even its origin), but having doubts about the nuclear explanation of the Tunguska event, made nonetheless a half-step toward its acceptance. Of course, he thought the TSB was a natural space body – an asteroid or the core of a comet. But how it exploded in the air is a separate question worthy of special consideration. Perhaps it was a thermal explosion? Why not? Such a hypothesis exists. But whatever was the cause of the explosion, this space body did definitely blow up – and such a powerful explosion, even a nonnuclear one, must have formed a fiery ball composed of plasma of high-temperature ionized gas. The fiery ball having expanded, its charged particles of opposite charges began to separate and move along the lines of force of the geomagnetic field. It was this motion (an electric current, in essence) that produced the geomagnetic storm.[13] However, Kim Ivanov proved that this model could not explain the duration of the effect. The TSB had exploded in the lower atmosphere, at a height less than 10 km, where high-temperature plasma can exist only several minutes before it recombines.[14]

But Ivanov himself, trying to exorcize from his calculations the evil spirit of nuclear reactions, created a very unconvincing model of the Tunguska geomagnetic effect. He believed that it could have been due to the thermal ionization of the ionosphere. Yes, if some volume of the rarified air of the ionosphere (which is, of course, already ionized by solar radiation) is heated up to the temperature of 6,000–7,000°C it would be additionally ionized. But what could have raised the temperature of the air so much? According to Kim Ivanov, it was the blast wave of the Tunguska explosion that had such a high temperature and therefore must have heated the ionospheric air. Alexey Zolotov did, however, demonstrate – mathematically and by referring to direct measurements from nuclear tests – that the Tunguska blast wave could not be so hot. In fact, even the blast wave of a powerful thermonuclear explosion has the temperature of 6,000°C at a distance of 1.5 km from the center of the explosion. And its temperature decreases very swiftly with distance. Thus, in the ionosphere the temperature of the blast wave of the Tunguska explosion would not have exceeded 200°C – which is absolutely insufficient for the thermal ionization.[15]

There is, by the way, one more reason that prevents us from accepting the blast wave theory as a satisfactory explanation of the Tunguska geomagnetic storm. All specialists agree that the artificial geomagnetic effects, discovered in the nuclear tests of 1958, were very similar to that recorded in 1908. The shapes of the curves, the relative durations, and the amplitudes of various phases are practically the same. So, Victor Zhuravlev drew the attention of the Tunguska research community to a very simple error that had been made by the supporters of the blast wave hypothesis.

As it follows from the models of Ivanov's and Kovalevsky's, both hard radiation and the blast wave could have led to the same result, that is, to the local geomagnetic effects. Well, let's accept for a while that the Tunguska explosion was not accompanied by hard radiation and the Tunguska geomagnetic storm was produced by nothing but its blast wave. But then, it means that after a *nuclear* explosion *two* geomagnetic effects would have been produced. The first generated by the hard radiation and the second by the blast wave. Since the velocity of propagation of hard radiation exceeds that of the blast wave by many thousands of times, the interval between them would have been about 5 min. Why, then, did the high-altitude nuclear explosions in the atmosphere produce only *one* geomagnetic storm from the hard radiation of the fiery ball? Where is the second from the blast wave?

Can we suppose that the blast wave of a high-altitude nuclear explosion traveled through the ionosphere not disturbing the geomagnetic field, whereas the same wave from the Tunguska explosion did disturb it? No, we cannot. If a blast wave could have produced the local geomagnetic effect, the high-altitude nuclear tests would have recorded "paired" geomagnetic storms – from the hard radiation of the fiery ball and from the blast wave. Since there is no evidence of this, it means that a blast wave cannot produce such an effect. This is impossible theoretically and was never found in experiment. It is only the hard radiation of the fiery ball that can produce the local geomagnetic effect.

Nonetheless, great pains were taken to explain the Tunguska geomagnetic storm, both inside and outside the ITEG, while not referring to the nuclear model of this event. Two founding fathers of the ITEG – Victor Zhuravlev and Valentin Demin – demonstrated that such attempts were doomed to failure.[16] Again, it was Alexey Zolotov who called a spade a spade. In the monograph *Problem of the Tunguska Catastrophe of 1908*, he developed a detailed quantitative theory of an artificial magnetic storm.[17] According to this theory, the main phase of the local geomagnetic effect after a nuclear explosion arises due to fast electrons emitted by its fiery ball and caught in the geomagnetic trap – the layer of the terrestrial magnetosphere, inside which the configuration of magnetic lines of force prevents charged particles from leaving it. The sequence of events may vary, depending on the altitude of the explosion. However, Zolotov has showed conclusively that all possible schemes of the geomagnetic effect are based on nuclear reactions only. No contribution from a blast wave is needed to explain it.

Does the "nuclear explanation" of the Tunguska geomagnetic effect have any weak points? Or does this model explain every detail perfectly? Yes, it has some weak points. The first obstacle that Zhuravlev, Demin, and Zolotov faced when developing the nuclear model proved to be the time lag between the moment of the Tunguska explosion and the start of the geomagnetic storm. Kim Ivanov estimated its duration as some 2 min. As for the high-altitude nuclear explosions over Johnston Island, there was no time lag at all – both on August 1 (the explosion magnitude 3.8 Mt, the height 78 km) and on August 12, 1958 (the same magnitude, the height 42 km). The first phase of the geomagnetic effect started immediately after the explosions, the delay being less than one second. Assuming that the velocity of the blast wave of the Tunguska explosion was transonic (340 meters per second) and the lower boundary of the ionosphere was located at 80 kilometers over Earth, Ivanov determined that the blast wave must have reached this boundary in about four minutes.

As these figures were of the same order, Kim Ivanov decided that it had been the blast wave that had produced the Tunguska geomagnetic storm. He completely agreed, however, that when a thermonuclear bomb exploded in the upper atmosphere, the geomagnetic disturbance was due to the hard radiation from the explosion. That is why there was no time lag between the moments of the explosions and the beginnings of the geomagnetic storms during the nuclear tests in the Pacific in 1958. Neutrons and gamma rays travel much faster than even a powerful blast wave.

In fact, the duration of the time lag between the moment of the Tunguska explosion and the start of the geomagnetic storm was then known with an accuracy of several minutes. It was therefore necessary to find out its *exact* value. But the only way to refine it would be determining, from other instrumental data, the exact moment of the Tunguska explosion itself.

It was Professor Ivan Pasechnik (1910–1988) who was asked by the academic Committee on Meteorites to take on this difficult task. Pasechnik was the leading Soviet specialist in monitoring foreign nuclear tests. He organized in the Soviet Union and supervised a net of observing stations that detected all nuclear explosions outside the USSR and measured their parameters. It was Pasechnik who persuaded his colleagues and government officials both in the Soviet Union and in the West that measuring instruments existing early in the 1960s could detect even the weakest nuclear explosions in every corner of the world. Thanks to this, the USSR, the United States, and the United Kingdom signed in 1963 the Partial Test Ban Treaty prohibiting nuclear tests in the atmosphere, in outer space, and under water.

One of the main methods of keeping track of nuclear explosions was by analyzing seismic waves of the explosions. The Tunguska explosion left records of its seismic waves on the bands of seismographs in Irkutsk, Tashkent, Tbilisi, and Jena – but only the Irkutsk and Jena seismograms exist today. Attempts were made to determine the exact moment of the Tunguska explosion from this seismic data, first by director of the Irkutsk Magnetographic and Meteorological Observatory, Arkady Voznesensky. He arrived at the figures 0 h 17 min 11 s GMT, but Voznesensky in his calculations used the "average" velocity of seismic waves known at that time, which made his result not too precise. Fortunately, in 1986, Russian geophysicists managed to measure the velocity of seismic waves along paths that practically coincided with the paths of those waves that had been recorded during the earthquake produced by the Tunguska explosion. And Professor Pasechnik used these data in his calculations. It turned out that the Tunguska explosion had occurred between 0 h 13 min 30 s and 0 h 13 min 40 s GMT.[18]

Now this is important, because we know that the Tunguska geomagnetic storm started at 0 h 20 min 12 s GMT. Therefore, the time lag was as long as 6 min 23 s. When we also consider that the blast wave of the Tunguska explosion took some 10 seconds to reach Earth's surface (obviously, the earthquake could not have started earlier), it means that the time lag was in fact about 6.5 minutes. And what of it the reader will ask? Well, this figure refutes the blast wave model for the Tunguska geomagnetic storm. With such a time lag, the speed of the blast wave that would have been needed in the ionosphere to produce a magnetic disturbance would have been 200 meters per second – much too low. The velocity of sound waves is about 330 meters per second, and no blast wave can travel below that speed.

So how did the time lag originate? In the theories of Ivanov's and Kovalevsky's it fits naturally. This is the time the blast wave had to reach the ionosphere. But the "nuclear" model of the geomagnetic effect did not need any time lag. Hard radiation propagates much faster than any blast wave, and it would have reached the ionosphere in a split second. This is why Alexey Zolotov tried to prove that there had been no real time lag between the explosion and the geomagnetic effect – it must have arisen, he said, in calculations due to the low precision of initial data. But Professor Pasechnik has convincingly proved that this was not the case; the time lag was for real and it was rather large. So where do we go from here?

Victor Zhuravlev, pondering this problem, noted an important detail: the fiery ball of the Tunguska explosion was usually thought of as stationary. It had to emit hard radiation but not to move.

Reality is different. The fiery ball of a nuclear explosion that occurs at a height of several kilometers almost immediately starts to rise into the stratosphere – just because it is lighter than air. And its ascent lasts 6–10 min.

This relatively slow motion of the fiery ball has to be the cause of a time lag. Only after reaching an altitude where the air density is low enough can the hard radiation of the fiery ball influence the ionosphere and produce a local geomagnetic effect. Since a store of radioactive substances in the fiery ball of a nuclear explosion is very large, the artificial geomagnetic storm can last one hour or more.

Thus, it seems that for 6 min 30 s after the Tunguska explosion its fiery ball was rising and only then the upper atmosphere felt the influence of its hard radiation. The concentration of electrons and ions in the ionosphere over the Great Hollow sharply increased. At that time a magnetic wave moved toward Irkutsk.[19] The result? The intensity of the geomagnetic field jumped, and this jump was detected by magnetometers of Voznesensky's Observatory.

In 2003, speaking in Moscow at "The 95th Anniversary of the Tunguska Problem" conference, Kim Ivanov agreed that the blast wave in itself could not have produced the geomagnetic effect. Additional ionization of the ionosphere over the place of the explosion was necessary for that. "The source of this additional ionization remains unknown," he said. It appears that after many years of investigations and discussions, the opinions of Tunguska researchers on the origin of the local geomagnetic storm – if not on the origin of the TSB – had drawn nearer.

True, the "additional ionization" does not necessarily imply a "nuclear explosion." The nuclear model of the geomagnetic effect just meets one more difficulty. The Tunguska local geomagnetic storm was, paradoxically, "somewhat too strong" and "somewhat too long" to be regarded as the final proof of the nuclear hypothesis of the Tunguska explosion.

How to explain this peculiarity? Victor Zhuravlev and Alexey Dmitriev suggested that the plasma cloud (without which no model of the regional geomagnetic effect would work) did not originate at the moment of the explosion. Instead, it came to the atmosphere of Earth as a "plasmoid" generated by the Sun. It was the American physicist Winston H. Bostick (1916–1991) who coined the term "plasmoid" in 1956, implying a coherent structure consisting of plasma within a magnetic field and able to exist for some time outside of the source that generated it.[20] Such structures arise, for example, when plasma is injected into a vacuum chamber in which a strong magnetic field exists. But the lifetime of these artificial plasmoids is rather short. As for the TSB, it could be, according to Zhuravlev and Dmitriev's opinion, a huge and stable natural plasmoid shaped as a spindle-like "magnetic bottle" and surrounded by an external magnetosphere.

Recombining over the Great Hollow, protons and electrons of the plasma cloud generated hard radiation, after which the process developed in the same manner as in the wake of a nuclear explosion. This radiation, in its turn, gave rise to a system of electric currents in the ionosphere that produced the regional geomagnetic effect. The amount of plasma in the "magnetic bottle" had to be great enough to maintain this system of currents for about five hours.[21]

Trying to calculate the strength of magnetic field for their model, Zhuravlev and Dmitriev have however obtained an unbelievably high figure: 16 teslas. Such a field would be stronger than the terrestrial magnetic field by about half a million times. Even though fields of this order of intensity have been produced in some terrestrial laboratories – with the help of superconducting solenoids – they have never been detected on the Sun. It seems therefore that attempting to introduce into the Tunguska problem a new "natural" hypothesis for the TSB origin, Zhuravlev and Dmitriev have instead built a novel version of its "artificial" model, something like a starship with, figuratively speaking, a "plasma-magnetic engine." For a purely natural object, the intensity of the magnetic field inside the hypothetical plasmoid would have been much too high. Besides, the idea itself bore little if any hard evidence – such objects have never been observed in the Solar System.

But whether or not this hypothesis can explain all the circumstances of the Tunguska event, it at least suggests that the TSB itself was the source of a strong magnetic field. And this supposition of Zhuravlev and Dmitriev's appears to have been confirmed not only by the local geomagnetic storm but also by a paleomagnetic anomaly in the soil of the Great Hollow.

Geophysicists have long been aware that many igneous rocks were magnetized when they formed. That is when hot liquid magma cools. More exactly, it is ferromagnetic minerals making up the rocks (especially, magnetite and hematite) that become, under such conditions, permanently magnetized. Usually the directions of these residual magnetizations are parallel to the direction of the geomagnetic field that existed at the time of their formation. When deposited in water basins, the magnetized minerals do also tend to align themselves along the lines of force of this field.

Although paleomagnetic research began to develop only after World War II, it has become a mature field of science that has, in particular, greatly helped to establish the theory of continental drift. The natural remanent magnetization is well maintained in the rocks and may be measured with modern magnetometers. In 1971, Saulas Sidoras, a specialist in paleomagnetic geological prospecting, and the mathematician Alena Boyarkina asked an important question: Could the same cause that had produced the geomagnetic effect recorded at the Irkutsk Observatory also have affected the residual magnetization of soils in the Great Hollow? Their work led to the finding of the Tunguska paleomagnetic anomaly.

It was a long and painstaking investigation. From an area of 600 km^2, in friable deposits of the near-surface layer of the soil, the researchers took samples that were marked with arrows indicating direction to the northern magnetic pole. After that, by a conventional procedure, the strength and direction of the natural remanent magnetization were measured in the lab.

The finding from this research is that there exist in the Great Hollow two components of residual magnetization instead of the usual one. This is definitely strange because one of these components coincides with the direction of the expected geomagnetic field while the other does not. Around the Ostraya Mountain, at a distance of about 4 km from the epicenter along the first of Fast's TSB trajectory (according to which the TSB was flying to the west-northwest), the structure of the remanent magnetization looks the most chaotic. It was therefore here that the magnetic influence of the TSB was greatest. "It seems reasonable to suppose," wrote Sidoras and Boyarkina, "that this effect is due to the influence of a magnetic field whose direction was opposite to the normal geomagnetic field. Such a field could decrease the residual magnetization."[22] Closer examination of the paleomagnetic anomaly in the Great Hollow has shown that zones of equal residual magnetization exist around the Ostraya Mountain, extending to the northwest and then to the north. Outside these zones the residual magnetization of local soils does not differ from the background one.

Figure 7.2 shows how this anomaly looks. Computations carried out by Victor Zhuravlev have led to the conclusion that the surface paleomagnetic anomaly could be produced by the same source that generated the first phase of the local geomagnetic storm of June 30, 1908. To disrupt the residual magnetization around the Tunguska epicenter to the extent that was measured by Sidoras and Boyarkina, the magnetic field imposed on the site of the catastrophe must have been 50–60 times stronger than Earth's magnetic field. But if the source itself was at an altitude of several kilometers, the strength of the field at its source must have exceeded the strength of Earth's geomagnetic field by 500 times! In Irkutsk, that is, at a distance of 970 km from the Great Hollow, such a source could have produced the start of the geomagnetic effect that was recorded at the Irkutsk Observatory.

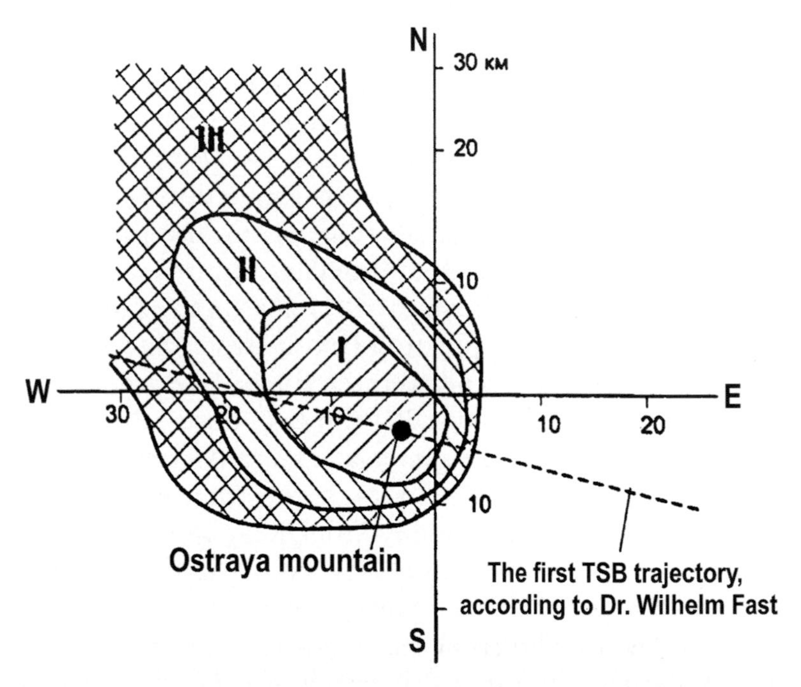

FIGURE 7.2. The area of the paleomagnetic anomaly testifying that the Tunguska space body was the source of a powerful magnetic field (*Source*: Vasilyev, N. V. *The Tunguska Meteorite: A Space Phenomenon of the Summer of 1908.* Moscow: Russkaya Panorama, 2004, p. 149.).

So the paleomagnetic anomaly and the local geomagnetic storm complement each other very well. But they are still not the strangest aspects of the magnetic trace of the Tunguska catastrophe. At least, they originated *after* the Tunguska explosion, being its results. But there also exists a third aspect of the magnetic trace of the event – the most enigmatic one, which may be called the "magnetic precursor" of the Tunguska phenomenon. We mean here the so-called "Weber effect."

By the irony of fate, the uncovering of this peculiar effect preceded discoveries about the local geomagnetic storm and the paleomagnetic anomaly. In the spring of 1959 two leaders of the ITEG – Gennady Plekhanov and Nikolay Vasilyev – were perusing scholarly journals dated back to the year 1908, looking for information that could have had anything to do with the Tunguska event. And suddenly they came across a short report published in the German *Astronomische Nachrichten* journal. It was entitled "Von Herrn Prof. Dr. L. Weber, Kiel, Physikalisches Institut der Universität, 1908 Juli 11." According to this report, Professor Weber, when working at a laboratory of Kiel University, Germany, observed from June 27 to June 30, 1908 a very unusual geomagnetic effect. "Throughout the last 14 days," he wrote, "the photographically recorded curves... did not demonstrate any disturbances that usually accompany aurorae. But I would like to note that several times, during many hours, were permanently observed small, regular, uninterrupted oscillations with an amplitude of two angular minutes and period of 3 min. These variations are not attributable to any known causes (say, to the disturbances arising from tramways in the city)."[23]

The variations were recorded three times. First, they started at 6 pm, June 27, and lasted 7 hours 30 minutes – until 1.30 am June 28. These oscillations recurred exactly at the same time interval on June 28–29 from 6 pm to 1.30 am. Next day, that is, June 29, they commenced at 8.30 pm and finally stopped at 1.30 am, June 30.[24] This time they lasted only 5 hours. Nikolay Vasilyev and his colleagues tried to find the originals of these magnetograms, but they had been destroyed during World War II.

As emphasized by one of the leading ITEG members, Boris Bidyukov, the beginning of the Weber effect falls upon that very day (June 27, 1908), when over Europe, and especially over Germany, became visible "optical precursors" of the Tunguska explosion – the peculiar light anomalies in the atmosphere. It finished in 16 min after the explosion. As far as we can judge, neither before nor subsequently were similar effects ever recorded. So, a chance coincidence of these events is highly improbable.

"The interval between these oscillations," Bidyukov writes, "was *24 hours exactly*, that is, one revolution of the Earth on its axis."[25] Perhaps, the only association that comes to mind in this connection is the idea of a satellite traveling in an elliptical orbit with a period of 24 hours and its closest point over Germany. If such a satellite was the source of a powerful magnetic field it could have influenced Professor Weber's magnetometer. We will later consider a complicated theory, recently developed by a group of Russian scientists, connecting the hypothetical "Tunguska comet" with the Weber effect. However, we have to agree with Professor Weber that these oscillations cannot be attributed to any known natural causes.

Now, in which direction does the "third Tunguska key" turn? One can say with confidence it does not point in the direction of a comet core or a stony meteorite. Rather, it points to a nuclear explosion, though the opinion of Alexey Zolotov and Victor Zhuravlev that the local geomagnetic storm is the final proof that the Tunguska explosion was nuclear should be viewed with some reservation. Anyway, the importance of this key should not be underestimated. Karl Popper believed that no hypothesis could be finally proved; it could only be "not falsified." In other words, a lot of evidence in favor of a hypothetical model does not mean it is entirely vindicated, whereas a single piece of evidence *against* it does refute the hypothesis. From this viewpoint, even if the magnetic traces of the Tunguska event have not fully established the correctness of the nuclear model, they at least may be considered as convincing evidence against the "standard" cometary-meteorite model. Neither the core of a comet nor a stony meteorite could have produced the local geomagnetic storm or have left a paleomagnetic anomaly at the epicenter of the explosion.

The favorite method of adherents to the meteoritic models of the Tunguska phenomenon is to declare any puzzling find a "chance cooccurrence." But in this case it does not work. The geomagnetic effect of June 30, 1908, differed radically from usual solar geomagnetic storms, being at the same time very similar to those geomagnetic disturbances that are produced by nuclear explosions in the atmosphere. And besides, would a global or near-global solar geomagnetic storm affect the residual magnetization just in the Great Hollow at the epicenter of the Tunguska explosion? Certainly not.

Notes and References

1. Strictly speaking, a gamma is 1/100,000 of an oersted, the unit of measurement of the magnetic field induction, but this is not important for our considerations. When we are dealing with magnetic fields in the vacuum or in a very rarified air, the difference between magnetic induction and magnetic field intensity becomes negligible.
2. See Matsushita, S. On artificial geomagnetic and ionospheric storms associated with high-altitude explosions. – *Journal of Geophysical Research*, 1959, Vol. 64, No. 9; Mason, R. G., and Vitousek, M. J. Some geomagnetic phenomena associated with nuclear explosions. – *Nature*, 1959, Vol. 184, No. 4688.
3. See Leypunsky, O. I. On the possible magnetic effect of high-altitude explosions of atomic bombs. – *Zhurnal Eksperimentalnoy i Teoreticheskoy Fiziki*, 1960, Vol. 38, No. 1 (in Russian).
4. Kalashnikov, A. G. On observation of the magnetic meteor effect by the induction method. – *Reports of the USSR Academy of Sciences*, 1949, Vol. 66, No. 3; Kalashnikov, A. G. Magnetic meteor effect. – *Reports of the USSR Academy of Sciences, Geophysical Series*, 1952, No. 6 (in Russian).
5. See Zhuravlev, V. K., and Zigel, F. Y. *The Tunguska Miracle: History of Investigations of the Tunguska Meteorite*. Ekaterinburg: Basko, p. 52 (in Russian).
6. Here is a more detailed description of the Tunguska geomagnetic effect, with some figures. The first entry led to an increase of the horizontal component of the geomagnetic field (H) for 4 gammas. Its second phase started at 0 h 22 min GMT with a new increase in the H magnitude. In the course of 18 min, it rose for 20 gammas more. For the next 14 min, the H component remained at the same level, after which, at 0 h 36 min GMT, the phase of fall began. During 1 h 41 min, the H component's value decreased by 67 gammas. The last phase started at about 2 h 17 min GMT and lasted some 3 h, until 5 h 20 min GMT (or 12 h 20 min, local time). The vertical component of the geomagnetic field (Z) did also change, although it returned to its usual value 2 h earlier

than the H component – at 3 h 20 min GMT. The magnetograms seemed not to show any change of the magnetic declination. But paying due attention to the usual daily variation of the geomagnetic field, Kim Ivanov and another Russian geophysicist, V. I. Afanasieva, succeeded in discovering alterations in the magnetic declination D as well. It turned out that the plane of the magnetic meridian had deviated by 10 angular minutes to the west and this deviation persisted during 5 to 6 hours.

7. See Zhuravlev, V. K. The geomagnetic effect of the Tunguska explosion and the technogeneous hypothesis of the TSB origin. – *RIAP Bulletin*, 1998, Vol. 4, No. 1–2, p. 9.
8. Zhuravlev, V. K. op. cit., p. 5.
9. Ivanov, K. G. Geomagnetic effects that were observed at the Irkutsk Magnetographic Observatory after the explosion of the Tunguska meteorite. – *Meteoritika*, Vol. 21, 1961 (in Russian).
10. Plekhanov, G. F., Kovalevsky, A. F., Zhuravlev, V. K., Vasilyev, N. V. On the geomagnetic effect of the Tunguska meteorite explosion. – *Proceedings of Institutions of Higher Educations. Physics*. 1960, No. 2.
11. Kovalevsky, A. F. The magnetic effect of the explosion of the Tunguska Meteorite. – *The Problem of the Tunguska Meteorite*. Tomsk: University Publishing House, 1963, p. 192 (in Russian).
12. Idlis, G. M., and Karyagina, Z. V. On the cometary nature of the Tunguska meteorite. – *Meteoritika*, Vol. 21, 1961 (in Russian).
13. Obashev, S. O. On the geomagnetic effect of the Tunguska meteorite. – *Meteoritika*, Vol. 21, 1961 (in Russian).
14. Ivanov, K. G. The geomagnetic effect of the Tunguska fall. – *Meteoritika*, Vol. 24, 1964 (in Russian).
15. Zolotov, A.V. *The Problem of the Tunguska Catastrophe of 1908*. Minsk: Nauka i Tekhnika, 1969, pp. 161–168 (in Russian).
16. See Zhuravlev, V. K. On the interpretation of the geomagnetic effect of 1908. – *The Problem of the Tunguska Meteorite*. Tomsk: University Publishing House, 1963 (in Russian); Zhuravlev, V. K., Demin, D. V., Demina, L. N. On the mechanism of the magnetic effect of the Tunguska meteorite. – *The Problem of the Tunguska Meteorite*. Vol. 2, Tomsk: University Publishing House, 1967 (in Russian).
17. See Zolotov, A.V. *The Problem of the Tunguska Catastrophe of 1908*. Minsk: Nauka i Tekhnika, 1969, pp. 155–191 (in Russian).
18. See Pasechnik, I. P. Refinement of the moment of explosion of the Tunguska meteorite from the seismic data. – *Cosmic Matter and the Earth*. Novosibirsk: Nauka, 1986, p. 66 (in Russian).
19. The so-called Alfvén wave – a traveling oscillation of the ions and the magnetic field.

20. See Bostick, W. H. Experimental study of ionized matter projected across a magnetic field. – *Physical Review*, 1956, Vol. 104, No. 2; Bostick, W. H. Experimental study of plasmoids. – *Physical Review*, 1957, Vol. 106, No. 2.
21. See Dmitriev, A. N., and Zhuravlev, V. K. *The Tunguska Phenomenon of 1908 as a Kind of Cosmic Connections Between the Sun and the Earth.* Novosibirsk: IGIG SO AN SSSR, 1984, pp. 125–127 (in Russian).
22. Sidoras, S. D., and Boyarkina, A. P. Results of paleomagnetic investigations in the region of the Tunguska meteorite fall. – *Problems of Meteoritics*. Tomsk: University Publishing House, 1976 (in Russian). See also Vasilyev, N. V. *The Tunguska Meteorite: A Space Phenomenon of the Summer of 1908.* Moscow: Russkaya Panorama, 2004, p. 149 (in Russian).
23. *Astronomische Nachrichten* Journal, 1908, Vol. 178, No. 4262, p. 239; see also: *Nature*, 1908, July 30, p. 305.
24. When recording the oscillations, Weber most probably used middle European time which differs from Greenwich time for an hour. We know that the Tunguska explosion occurred at 0 h 14 min GMT; therefore, the variations of the magnetic needle in Kiel stopped 16 min after the moment of the explosion.
25. Bidiukov, B. F. The "Weber Effect" and anomalous luminous phenomena in the Earth's atmosphere in the period of the Tunguska event of 1908. – *RIAP Bulletin*, 2006, Vol. 10, No. 2, p. 13.

8. Significant Details for the Big Picture

In the last chapters we considered three large keys to unlock the Tunguska mystery – the "mechanical," "thermal," and "magnetic." And now we must look at five smaller keys discovered in the course of Tunguska investigations. Practically every time such a new key emerged, the investigators were highly surprised. These are the supposed material remnants of the Tunguska space body (TSB): the "material" key, the "botanic" key (the superfast restoration of the Tunguska forest), and the "genetic" key (mutations in trees and other living things). But also there are fluctuations of radioactivity (the "radioactive" key), and last but not least, evidence of the ionizing radiation that had probably affected the Tunguska soil in 1908 (the "thermoluminescent" key).

Although the word "large" is a synonym of "primary" and "important," the word "smaller" does not necessarily mean "unessential" or "secondary." Quite the contrary, the first trace from the group of "smaller Tunguska traces" – possible material remnants of the TSB – is probably the most important of all potential traces of this enigmatic event. Factually, it is only these remnants that may be called its direct trace; any other piece of evidence, even one so massive as the radially leveled forest over an area of 2,150 km^2, is only indirectly connected with the TSB.

Professor Nikolay Vasilyev, when summing up the experience of his 40-year Tunguska studies, said: "The main paradox of the current situation is that no cosmic substance has been found as yet that could be reliably identified as the substance of the Tunguska meteorite."[1] Does it mean that this substance had mysteriously left our world, and we should give up all attempts to retrieve it? Of course not. In this case, we would simply have abandoned any hope of solving the Tunguska problem. Indirect traces, even important and informative, can at best outline a border between the possible and impossible, rather than give the final answer to the question of the nature of the Tunguska phenomenon. To find out

V. Rubtsov, *The Tunguska Mystery*, Astronomers' Universe,

DOI 10.1007/978-0-387-76574-7_8,

what was the nature and origin of the TSB, we must find its material remnants; otherwise this mystery will remain unsolved forever.

It's a pity that neither spacecraft debris nor meteorite pieces have been found, despite long and intensive searching. Why? Did the Tunguska researchers use methods that were not sufficiently sensitive? As Professor Vasilyev has written, several varieties of space dust that continually fall onto Earth's surface have been discovered. Of course, if these methods were sensitive enough to find traces of dust from space they should have been good enough to discover remnants of a huge space body dispersed in the soil and peat.[2] Does it mean therefore that there are none?

Tunguska researchers have always believed that the TSB substance is still preserved somewhere in the taiga and may be found. The only exception is probably Lincoln La Paz and his antimatter hypothesis, according to which the Tunguska meteorite was completely annihilated in the terrestrial atmosphere. But this is an extreme viewpoint. A piece of antimatter would hardly have penetrated Earth's atmosphere so deeply – it would have been annihilated at a higher altitude. Also, as astronomer Vitaly Bronshten has demonstrated, small bodies of antimatter could not even traverse the Solar System without being destroyed when interacting with interplanetary gas.[3]

Of course, specialists in meteoritics looked for more normal matter. First, they tried to find in the Great Hollow large pieces of meteoritic iron (Leonid Kulik) and then small metallic spherules (Kirill Florensky). Leonid Kulik was absolutely sure that the TSB had consisted of nickelous iron, which was perfectly reasonable because all large meteorites found on Earth's surface are blocks of iron. The largest known mass of cosmic iron, the Hoba meteorite that landed near Grootfontein in northern Namibia, weighs about 60 tons. It collided with Earth approximately 80,000 years ago.

Some researchers used to speak ironically about Leonid Kulik's bent for the iron-meteorite model of the TSB, but in fact he knew well that other types of meteorites had little if any chance to reach Earth's surface. Stony meteorites are split into many pieces in the upper layers of the atmosphere and their small pieces could not have produced such devastation in the taiga. But not every rational hypothesis in science turns out to be correct. Yet even though Kulik had failed in his search, Florensky became certain that the metallic spherules found in the taiga in 1961 were the TSB substance. At least, so he said.

In the early years of the space era one could assume that the main mass of a large piece of cosmic iron would burn up during its flight through the atmosphere. The laws of such flights had been scantily investigated. But specialists soon proved that an iron meteorite would leave a pronounced trace in the soil. So given that the TSB was an iron meteorite, about 90% of its mass would have fallen at the central area of the Great Hollow and only 10% would have dissipated in the upper layers of the atmosphere.[4]

The ITEG tried every way to find the TSB substance, and some spherules of meteoritic iron were found. But to prove or disprove that these spherules have something to do with the TSB, they had to be reliably dated. It was the Siberian botanist Yury Lvov (1932–1994) who saw how this could be done simply and effectively. One of various mosses that grow on Siberian peat bogs is the so-called golden sphagnum. This plant has two characteristics that proved to be very useful for Tunguska studies. First, it obtains mineral nutrition not from the soil but from atmospheric substances, absorbing fine particles including falling space dust. It also grows at a steady rate, making it possible to determine the age of its yearly layers with high precision. Consequently, a vertical column of peat shows the past history of space dust falls for many tens and sometimes hundreds of years.

Lvov's method had been tested on peat bogs both in Siberia and in European Russia. Everywhere it proved to be effective and could therefore be used at Tunguska, although technically it turned out not to be that easy. Since outside the taiga the samples collected could have been contaminated with industrial dust, this research was being carried out in a forest. Among all research programs carried out by the ITEG, the "Peat" program was probably the most laborious. Samples have been dug up over an area of more than 14,000 km^2, the number of peat columns exceeding 1,000. The peat layers were burned in a muffle furnace and exposed to strong acids. What remained was scanned by a microscope in the search for fused microscopic spherules.[5]

Both silicate and metallic spherules, some 100 microns in diameter, were discovered in the peat, including the layer dated 1908. Significantly, in several places the number of spherules in the 1908 layer was much greater than in the lower and upper peat layers. But strangely enough, the concentration of the particles extracted from Tunguska soil and peat did not match other traces of the catastrophe of 1908, such as the borders of the area of leveled forest, or the light burn, or the direction of the TSB flight before the explosion. But if this dust had had anything to do with the TSB, this association would have been practically inevitable. And besides, the number of these spherules was simply too small even for a comet core, to say nothing of a huge stony meteorite. When extrapolating the data obtained, the overall mass of space matter spread over the Great Hollow in 1908 was somewhere between 200 kilograms and one ton. But, according to the well-justified estimation of Academician Vasily Fesenkov, the mass of the hypothetical Tunguska comet could not have been less than a million tons. A powerful explosion of the comet core entering Earth's atmosphere could have happened only if both its mass and its velocity had been very considerable. And now – 200 kg... Strange indeed. So most probably the main part of these microscopic spherules was due to the usual background fall of extraterrestrial matter.

Well, the *main* part, perhaps. But does it mean that there is among this space dust not a single microscopic particle of the Tunguska body? Deposits of usual microscopic space dust cover the surface of our planet unevenly, as do the radioactive fallouts after nuclear explosions in the atmosphere. By analogy, one can assume that after the Tunguska explosion there must have formed on the surface a patchy structure, within which there may be found spots more or less enriched with the TSB substance. Therefore, the researcher must not be nervous of different results of analyses even in two neighboring places. Statistical data are definitely important, but information obtained at some specific points may also hint at the nature of the Tunguska "meteorite."

True, at first the patchy character of the fall of space dust had somewhat embarrassed Tunguska researchers. But experienced radiochemists (i.e., specialists in the chemistry of radioactive materials) Sokrat Golenetsky and Vitaly Stepanok, who worked on the Tunguska problem together with Alexey Zolotov at a geophysical institute in the Russian city of Tver, succeeded in transforming the patchy character of cosmic matter into a new opportunity. If the cosmic matter is distributed over the Great Hollow nonuniformly, let's look for individual locations contaminated by the TSB substance. "Empty" columns of peat or soil may safely be ignored, whereas "rich" columns should be studied in detail.

If in addition to the very powerful and high-altitude main explosion, there were at Tunguska several more low-altitude explosions, then some places of the Great Hollow could be contaminated by the TSB substance.[6] Of course, the microscopic silicate spherules were too few to be considered as the main mass of the Tunguska comet. (Golenetsky and Stepanok generally shared the cometary hypothesis to explain the TSB.) However, a great part of its substance could have been dispersed in the air as an aerosol or simply vaporized. That is why attention had to be concentrated on the anomalies in the elements in the soil and peat, not on the spherules.

Even though they supported the cometary hypothesis, Golenetsky and Stepanok knew that it would have been premature to consider this as the final solution of the Tunguska problem. Aerial photographs taken by Leonid Kulik in 1938 demonstrated that there were in the Great Hollow several local centers of forest leveling. So the soil and peat in these centers might be enriched with the TSB substance, and finding it could help to solve the Tunguska problem. Sokrat Golenetsky had personally collected in one of these centers – near the Suslov's crater – samples of moss and peat from various depths. Two other columns of peat were taken at some distance from this place. As it turned out, in the "catastrophic" layer (dated 1908) and the neighboring peat layers of Column 1, concentration of certain chemical elements, such as sodium, potassium, chromium, zinc, bromine, rubidium, barium, mercury, and gold, was unusually high. High concentrations of zinc (an element of limited occurrence in meteorites), bromine, gold, and mercury looked very enigmatic, especially that of mercury – since when the peat was ashed for investigation, this element must have actively evaporated and therefore its *initial* concentration must have been still higher.

Two other peat columns did not demonstrate evident anomalies. The "patchy pattern" of the cosmic matter falls showed itself once again, but judging from the first peat column the composition of the TSB substance seemed to differ radically from all known types of iron or stony meteorites. What alternative might have been found? Perhaps a comet core, but first Golenetsky and Stepanok put forward a more original idea: it was an archaic space body, older than usual comets and carbonaceous chondrites, which had survived until now from an early epoch of the Solar System's formation.[7] To erect a new astronomical hypothesis on the basis of a single column of Siberian peat would be, according to all scientific standards, more than risky, and Alexey Zolotov expressed his negative opinion on this hypothesis very bluntly. One cannot say his criticism was unjustified. But the creativeness of Golenetsky and Stepanok does deserve respect.

This author happened to be a witness, if not a participant, of this dispute. It was hot indeed and, as sometimes happens in scholarly discussions, it soon went beyond a peaceful talk. Sokrat Golenetsky broke off friendly relations with Alexey Zolotov and left Tver. Subsequently he worked hard in the Chernobyl zone, examining the consequences of the greatest nuclear energy disaster in history, which probably precipitated his untimely death in 1996. But until the very last days of his life, Golenetsky remained active in Tunguska studies. With time, both his and Stepanok's positions in the Tunguska problem shifted from an archaic space body from the protoplanet cloud to a normal comet core. It was their research results that drew the attention of Dr. Evgeny Kolesnikov, a geochemist at Moscow University, and gave him the idea to check their validity, applying more sophisticated analytical methods.

At first, Kolesnikov verified that in the Tunguska peat layer dated 1908 concentrations of sodium, zinc, gold, and some other elements had really been increased. That is, Golenetsky and Stepanok were right. He also found that the concentration of iridium (a very hard and dense metal from the platinum group) in the 1908 layer was abnormally high. Iridium is very rare on Earth's surface but relatively common in meteorites. And having analyzed his data, Evgeny Kolesnikov concluded that the TSB had been a comet's core.[8]

Unfortunately, attempts to verify his conclusion when looking for traces of the Tunguska-related iridium anomaly in Antarctica and Greenland failed.[9] Yet, if a giant stony meteorite or a comet core had disintegrated over central Siberia in 1908, noticeable quantities of this metal must have remained in the pure ice of these distant regions of our planet. A deadlock? But are the soil and peat the sole possible repositories of microscopic TSB remnants? What else could harbor significant evidence?

Trees, of course! Since they were standing in the Great Hollow in 1908, scattered particles of the enigmatic space body could remain in them, too. Although it would be difficult to determine the age of those particles that have stuck in tree trunks and branches, there still remains tree resin. In the early 1990s specialists from Bologna University took resin samples in the central area of the Great Hollow to examine in Italy.[10] With a scanning electron microscope (in which the surface of a sample is scanned by a beam of electrons that are reflected to form an image) they found in separate layers of the resin a number of microscopic particles and determined their chemical composition. The Italian scientists examined more than 7,000 particles, each a few microns across. And they also found the same chemical elements that had been discovered by Golenetsky, Stepanok, and Kolesnikov in Tunguska peat. In particular (and especially), these included copper, zinc, gold, barium, and titanium. But also there were calcium, iron, silicon, and nickel. The Italian scientists paid their main attention to the latter group of elements. From the data they decided that these microscopic remnants were the remains of a small stony asteroid.

The final answer? Not yet, alas. It so happened that the Tunguska catastrophe occurred between major eruptions of two volcanoes: Ksudach on Kamchatka in 1907 and Katmai on the Aleutian Islands in 1912. These eruptions ejected into the atmosphere an enormous mass of volcanic ash. Early in 1908 Ksudach's ash fell even on Germany.[11] Consequently, as the resin layers containing enigmatic microscopic particles in Tunguska trees can be dated with an accuracy of 2–3 years, how can we be sure that these particles got there in 1908? Also, in 1980, Professor Claude Boutron of the Laboratory of Glaciology of the French National Center of Scientific Research discovered volcanic ash in Antarctic ice dated 1912 whose composition is very similar to that of particles found in the resin of Tunguska trees by the specialists from Bologna University. Whether the particles discovered by the Italian scientists were due to the Tunguska explosion or to the two volcanic eruptions remains unknown.[12]

The most systematic search for elemental anomalies in Tunguska soils and peats has been conducted by the ITEG people. It was after the ITEG-1 expedition of 1959 that the chemical composition of the samples taken at Tunguska was studied for the first time. The researchers had expected to find the usual meteoritic elements of iron, nickel, and cobalt. Instead, the spectral analysis demonstrated an increased concentration of some rare earths (lanthanum, ytterbium, cerium, and yttrium, which are designated in chemistry as lanthanides – from lanthanum, the first element of this series).[13] The concentration of such rare earth metals exceeded the norm by tens and even hundreds of times. Soon it turned out that the samples enriched by rare earths are found only around the epicenter and in the northwestern direction from it.[14] This chemical anomaly was spread through soils, plants, and peat, having a peak in the peat stratum dated 1908. So the TSB might have been composed of lanthanides.

Nevertheless, to avoid possible errors and to prove this supposition statistically, the ITEG started a special research program. It was necessary to find out if the rare earth anomaly was not connected with geochemical peculiarities of the region. The main attention was paid to the area lying in the west-northwestern direction from the epicenter, where, as John Anfinogenov had supposed, remnants of the TSB might have fallen. To carry out this work, members of an ITEG expedition cut a straight path 12 km in length through the taiga, running from the epicenter to the west-northwest through a peat bog that was subsequently named "Lvov's bog" after Dr. Yury Lvov. In the 1980s, having examined this place in detail, Lvov's pupil Emelyan Muldiyarov found that before the Tunguska explosion there had been at this place a normal forest, not a bog. This appears to be *the only place* at Tunguska where the landscape had changed drastically after the catastrophe. As such, it was definitely worth the researchers' special attention.

In an area 12 km long and 6 km wide, they took some 1,300 samples of soil and peat. After drying, milling, and sifting them, these samples were spectrally analyzed at a geological institute in Novosibirsk that was engaged in uranium ore prospecting and other nuclear-related work. Their measuring equipment, run by specialist Lidia Ilyina, could reliably determine the presence of 50 chemical elements, and from these they found 30 elements, including rare earths. And Ilyina noticed an astonishing fact: in some samples concentrations of yttrium and ytterbium were very close. In other samples there was plenty of ytterbium and no yttrium. But from the geological point of view this was simply impossible. In terrestrial rocks and minerals these two elements are inseparable, and the content of yttrium always exceeds the content of ytterbium by a factor of 10. As was mentioned above, meteorite specialists were not interested in rare earths, since these elements are far from typical for meteorites and comet cores.

Besides, this geochemical anomaly at Tunguska was not easily noticeable. But the ITEG included not only geologists and meteor specialists but also radio physicists. And one of the most important tasks that are solved by specialists in radiolocation is detecting a signal whose peak value is considerably lower than the level of the background noise. Radio physicists have developed sophisticated methods for extracting such signals from a chaos of radio waves. Dr. Dmitry Demin used this approach in Tunguska studies to create a special statistical method aimed at the search for "hidden anomalies" veiled by the surrounding "noise." Components of the TSB were considered as the "signal" and mundane chemical elements inherent in the soils and peats of the Great Hollow as the "noise."

To prove the strength of this method, ITEG researchers experimented with a simulated hidden anomaly. They took a map of nickel distribution in the Great Hollow and increased the figures, as if adding to the whole area 10 tons of this metal. An iron meteorite weighing just 100 tons (or a stony one weighing 1,000 tons) that had disintegrated over this area would have contained such an amount of nickel. At first glance nothing in the distribution of nickel in this area changed, but when the figures were processed on a computer the simulated anomaly was immediately detected.[15]

Having proved the effectiveness of Demin's method, it became possible to look for real hidden anomalies of distribution of chemical elements in the Great Hollow. If some element had showed a peculiar distribution associated with the epicenter or a probable TSB trajectory, this would have meant it had been part of the TSB. And after processing the results of the spectral analysis, the researchers obtained a significant result. They found that the maximum of ytterbium concentration was at a point near Ostraya Mountain where, according to John Anfinogenov, the remnants of the TSB must have reached Earth's surface. And the minimum of ytterbium fell on the "epifast," that is, the epicenter of the Tunguska explosion determined by Wilhelm Fast. Also, the straight line connecting these two points coincided with the first TSB trajectory calculated by Fast (see Figure 8.1.)

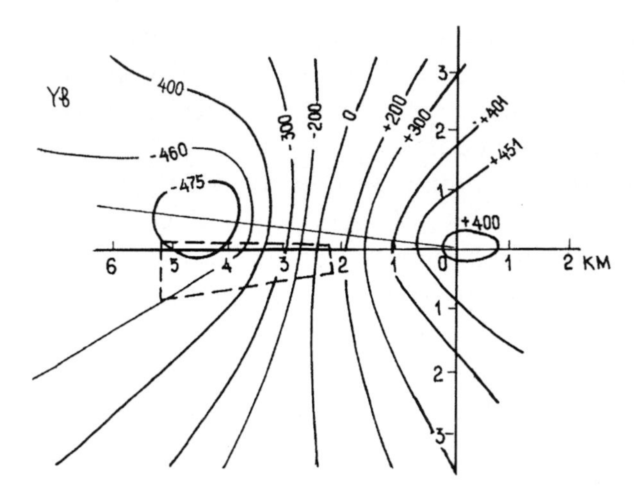

FIGURE 8.1. Pattern of ytterbium distribution at Tunguska following the projection of the TSB trajectory on the Great Hollow (*Source*: Zhuravlev, V. K., and Zigel, F. Y. *The Tunguska Miracle: History of Investigations of the Tunguska Meteorite*. Ekaterinburg: Basko, 1998, p. 110.).

Patterns of similar shapes have been formed in the Great Hollow for the surface distributions of lead, silver, and manganese, but for iron, nickel, cobalt, and chromium, the patterns of their distribution had no association with any special points or directions of the area of the leveled forest. These elements were therefore just natural components of the soil and rocks.

Here again the "negative" result seems almost more interesting that the "positive" one: calling a spade a spade (or, in Russian, calling a cat a cat), we should conclude that *typical meteoritic elements – iron, nickel, cobalt – have nothing to do with the Tunguska space body*.

But the "positive" result from this research is also worth attention. As the Siberian scientists state, from the 30 chemical elements discovered in the soils and peats of Tunguska, it is first of all ytterbium that can be reliably associated with the TSB. Also, possibly lanthanum, lead, silver, and manganese.[16] Certainly, with this composition, it could have been neither a meteorite nor a comet core.

Besides, let us not forget about the enigma of the rare earths' *ratio*. It looks very puzzling. Geochemists and geologists are well aware that in the presence of lanthanum there have to be cerium, neodymium, praseodymium, and other members of this family.

What is more, mutual ratios of their concentrations in rocks are fairly stable, fluctuating insignificantly. Not so at Tunguska.

In the 1980s, Dr. Sergey Dozmorov, a specialist in the chemistry of rare earths, who ran a chemical laboratory at a research institute in the Siberian city of Omsk, became interested in this enigma. He tested samples of soil, taken near Ostraya Mountain, for the presence of all lanthanides, not only of lanthanum, cerium, and ytterbium. Dozmorov discovered that, apart from ytterbium, these samples were enriched by thulium, europium, and terbium as well (these are also rare earth elements). And their ratio had been sharply disrupted. The contents of terbium exceeded the norm by 55 times, that of thulium by 130 times, that of europium by 150 times, and that of ytterbium by 800 times. Such things never happen in nature – only in special alloys. Even being a cautious scientist, and not a sensation-seeking journalist, Sergey Dozmorov had to conclude that:

> Together with the known data on the above-average barium content in the area of the Tunguska explosion, the results obtained may favor the most unusual composition for the TSB, namely the presence in the TSB of some systems that contained a superconducting high-temperature ceramic made on the basis of a combination of barium – a lanthanide – and copper. Such a ceramic keeps superconductivity up to the temperature of liquid nitrogen (–196°C) and can be used for constructing effective energy and information storage devices. Obviously, such a substance cannot be natural.[17]

Dozmorov was planning to continue and develop his research, but soon after obtaining this striking result he perished at night in his laboratory. Police investigators, who looked into this case, concluded that it was just an accident. Somehow the experienced chemist was poisoned by a toxic chemical compound. Such things happen. At that moment Sergey Dozmorov was 36 years old and was one of the leading Russian specialists in rare earth elements.

However, after this fatal accident the ITEG people did not give up. They continued to investigate the rare earths at Tunguska. During the expedition of 2001 a team guided by Dr. Victor Zhuravlev took from Lvov's bog a large column of peat. In Novosibirsk the samples were spectrally analyzed in three independent laboratories and it was found again that concentrations of some lanthanides (ytterbium, lanthanum, and yttrium) were considerably higher than normal.

Then the peat was examined through optical and electron microscopes, and Dr. Leonid Agafonov at the Institute of Geology of the Russian Academy of Sciences noticed several metallic particles that were, according to him, definitely artificial (see Figure 8.2). It was for the first time in the history of Tunguska investigations that someone had discovered microscopic artifacts in the peat layer dated 1908. And these are definitely not small pieces of Evenk teapots.

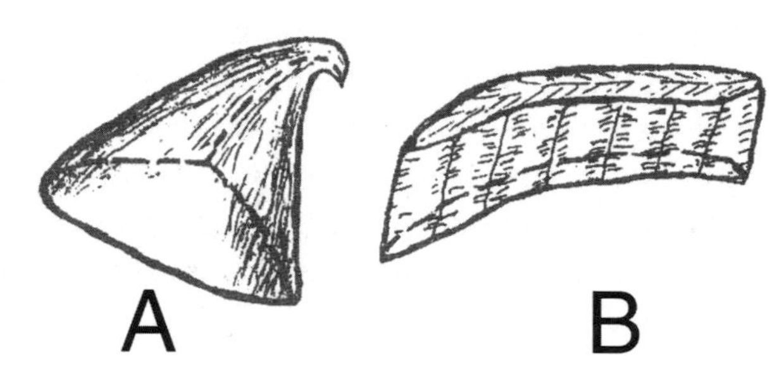

FIGURE 8.2. Peculiar microscopic artifacts discovered by Dr. Leonid Agafonov at the Institute of Geology of the Russian Academy of Sciences in the Tunguska peat layer dated 1908. The small trihedral pyramid **A** consists of pure titanium; the "shaving" **B** of aluminum (*Source*: Zhuravlev, V. K., Agafonov, L. V. Mineralogical and geochemical examination of the samples of soils taken in the area of the Tuguska bolide's disintegration. – *The Tunguska Phenomenon: Multifariousness of the Problem*. Novosibirsk: Agros, 2008, p. 151.).

The particles were curiously shaped and had an unusual chemical composition. There was a small trihedral pyramid with an edge of one-fourth of a millimeter consisting of pure titanium with some quantity of rhodium (a noble metal from the platinum group). A second particle looked like a bent microscopic plate (a "shaving") of about 250 microns in length. It consisted of aluminum with slight manganese and copper impurities. There were also found in these samples two small flattened balls of pure gold. As Dr. Zhuravlev noted in 2008, "We should not jump to conclusions from these findings. Yet we can probably hope to find in this area, near Ostraya Mountain, a larger remnant of the Tunguska space body. There seems to be at this area a 'geochemical halo' surrounding the place of its fall."[18]

In recent decades, Tunguska researchers have suggested as possible chemical constituents of the TSB a lot of various elements. These were aluminum, barium, bromine, calcium, carbon, cesium, cobalt, copper, gold, hafnium, iron, lanthanides (ytterbium, lanthanum, samarium, europium, thulium, terbium, cerium, dysprosium, gadolinium), lead, manganese, mercury, molybdenum, nickel, rubidium, silicon, silver, sodium, strontium, tantalum, tin, titanium, tungsten, zinc, and zirconium. A long list indeed! But only five elements in it – ytterbium, lanthanum, lead, silver, and manganese – have patterns of distribution in Tunguska soils and peats that follow the projection of the TSB trajectory on the Great Hollow, and only ytterbium follows this path strongly enough to be considered as the most probable main ingredient of the TSB substance.

An amazing outcome, one should note. In fact, there is nothing special in this chemical element ytterbium. This soft silvery-white rare earth metal, discovered in 1878, has at present very limited technical applications: it is used mainly for improving the hardness of stainless steel as well as in making high-power lasers. In the Solar System its occurrence is much rarer than in Earth's crust.

With such a peculiar composition, far from typical for normal meteorites, it is hardly surprising that the spectrum of theoretical interpretations of this data is so broad. Sokrat Golenetsky and Vitaly Stepanok saw in the TSB an archaic space body from an early epoch of the Solar System's formation, whereas Evgeny Kolesnikov believes it was a comet, and Giuseppe Longo and Menotti Galli consider it a stony asteroid. Each time these conclusions were well justified. Let's not forget, however, that the main elements, constituting all normal small cosmic bodies – iron, nickel, and cobalt – although discovered at Tunguska, do not display any correlations with the structure of the leveled forest area. This is curious indeed, since such correlations *must* have existed – *if* the TSB was such an ordinary space object. And if its chief chemical component was ytterbium, the nature of the TSB becomes still more incomprehensible. As far as we can judge, there are no known small space bodies in the Solar System consisting mainly of this element.

Let's remember that "ballistic" calculations, considered in Chapter 6, have also led to three equally well-justified hypotheses about the nature of the TSB: a comet, a stony asteroid, and an unknown space body. It seems that the "material" key to the gate of the Tunguska fortress turns freely in the same three directions, not stopping anywhere... What a maze! So where should we look for an exit from it? Probably it would be reasonable to pay attention to some other traces of the Tunguska phenomenon, rather biological than chemical, but also having a close relation to the question of the composition of the TSB.

Some years ago, Academician Nikolay Vasilyev, together with botanist Lyudmila Kukharskaya, tried to find out whether a watery extract from Tunguska soils, taken near the epicenter, would influence the process of the sprouting of pine and wheat seeds. It did influence them – and very positively, stimulating their germination. And what is more, it turned out that of all 35 chemical elements discovered in the Tunguska soil, only rare earths – lanthanum, ytterbium, and yttrium – had this "stimulating property."[19] Why is this so important? Because there exists one more enigma of Tunguska – the unusually fast restoration of the area in the aftermath of the catastrophe.

This mysterious phenomenon was discovered during the first academic expedition to Tunguska after World War II – in 1958, by Dr. Yury Emelyanov. Together with Dr. Valery Nekrasov, he examined the region thoroughly. Especially strange seemed the fact that even old trees, which had been burned by the light flash and seriously injured by the blast wave of the Tunguska explosion, did also accelerate their growth. From the viewpoint of forestry science this was incomprehensible. Even mosses in open marshy terrains started to grow much faster after 1908. Emelyanov and Nekrasov eventually concluded that this effect could not be explained by the improvement of environmental conditions for those trees that had survived the Tunguska catastrophe. Rather, it must have had to do with some stimulating substance that had dispersed over the Great Hollow after TSB's disintegration.

Why did scientists put forward this idea? First, because the boundary of the area of the superfast forest restoration was completely different from the boundaries of the zones of the wood fire and leveled trees (see Figure 8.3). The blast wave of the Tunguska explosion caused the major devastation in the southwestern and northeastern sectors of the Great Hollow, whereas trees grow unusually fast mainly in the opposite sectors – located to the northwest and southeast from the epicenter.[20] If this effect had had any relation to the ash fertilizers from incinerated vegetation or better light conditions in the devastated area, this certainly could not have happened. Besides, the axis of symmetry of the zone of the superfast forest restoration runs from the south-east to the north-west – coinciding with

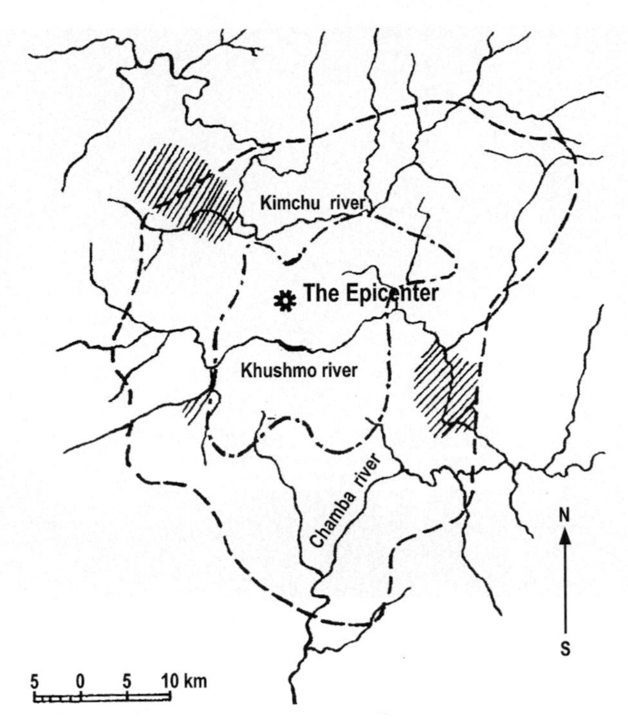

FIGURE 8.3. The hatched spots designate the areas in which trees, burned by the light flash and injured by the blast wave of the Tunguska explosion, grew at an abnormally fast rate (up to 36 times). This effect is incomprehensible from the viewpoint of forest science (*Source*: Vasilyev, N. V. *The Tunguska Meteorite: A Space Phenomenon of the Summer of 1908.* Moscow: Russkaya Panorama, 2004, p. 197.).

the "first TSB trajectory" determined by Wilhelm Fast. It is "under the trajectory" that this effect is most prominent. Here, before the catastrophe, diameters of larches had increased at about half a millimeter per year, whereas after this event, their average annual growth rate increased by an amazing 36 times than what was normal, reaching almost 2 cm (see Figure 8.4).

It is remarkable that there are near the epicenter some fairly large groves of pines and larches that have no signs of thermal burn or leveling. And these trees also grew abnormally fast after 1908. However, between the Kimchu and Moleshko rivers, where the forest was felled by the blast wave, no unusually swift wood restoration has been discovered. And finally, the scale of this effect goes far

FIGURE 8.4. A section of a larch that survived the 1908 disaster. Its rings after 1908 are noticeably wider than before (*Credit*: Vitaly Romeyko, Moscow, Russia.).

beyond the limits known to specialists in forestry. This is the *only* case when a forest suddenly began to grow so fast.

All these facts demonstrate that neither the forest leveling (that led to better light conditions in the taiga) nor the usual after-catastrophe fertilizers (wood ash) had anything to do with this enigmatic effect. Of course, they could contribute to it, but definitely they were not its main cause. But if the soil enrichment and more light may have only an indirect relation to this effect, what was its main cause? Which stimulant could affect so strongly the quality of the Tunguska trees?

Sokrat Golenetsky and Vitaly Stepanok thought it was cometary or "protoplanetary" substances that had fallen in the Great Hollow and enriched the soil with some microelements that turned out to be effective fertilizers.[21] To verify their hypothesis, they made a compound that reflected their ideas of the TSB's composition and conducted a series of experiments at the Research Institute of Land Reclamation, giving a top dressing of this compound to meadow grass, potatoes, and flax. They carried out the experiments in full accordance with the requirements of agronomy and achieved interesting results. The yield of potatoes rose by 30% and that of meadow grass by 20%. Good, but far from the growth acceleration by 36 times as occurred at Tunguska. Let's recall that before the Tunguska catastrophe the average width of the annual rings was only 0.2 mm, whereas after the explosion it reached in some places of the Great Hollow 1.8 cm.[22] Then, perhaps is what we have here a genetic mutation?

Golenetsky and Stepanok have waved the matter of mutations aside with some flippancy. "Attempts to explain the effect of superfast forest restoration by genetic mutations, allegedly produced by the 'hard radiation of the explosion,'" they wrote, "cannot be accepted seriously since all 'nuclear' hypotheses of the Tunguska explosion have been completely refuted."[23] This statement seems more emotional than rational, owing to the quarrel between Sokrat Golenetsky and Alexey Zolotov over the cosmochemical constructions built by Golenetsky on the foundation of one peat column, taken near the Suslov's crater. Both Golenetsky and Stepanok, as clever people and experienced specialists, were to understand that declaring the nuclear hypothesis "completely refuted" was an exaggeration. Besides, as we will see, genetic mutations at Tunguska do occur. This question has been studied by specialists for a long time and their final conclusion was in fact positive. Therefore, the abnormally fast restoration of the taiga could also be a genetic phenomenon.

But first, let us start from a basic question: what is mutation? In terms of modern genetics, a mutation is a change in a gene that alters the genetic message carried by that gene. Mutations may be lethal (resulting in a swift elimination of their carriers) or neutral (not affecting the further lot of living organisms). There are also point mutations that cause slight alterations of an organism's outer appearance, behavior, and so on. It is the point mutations that are the driving mechanism for changes by natural selection, which can lead to biological progress.

It was as far back as the early 1960s when the Commander of the Independent Tunguska Exploration Group, Gennady Plekhanov, understanding that an atomic explosion would have left too feeble radioactive traces to be detected after 50 years, attempted to find evidence in an indirect way. The hypothetical Tunguska radionuclides could have already decayed, but results of their influence on local plants might be preserved. Then at the Research Institute of Cytology and Genetics of the Siberian Branch of the USSR's Academy of Sciences, a group of scientists carried out experiments in which pine seeds were exposed to gamma radiation. Normally, a needle cluster of Siberian pine consists of two needles only. However, when a pine tree grows from a seed subjected to a small dose of gamma rays, there appear a considerable number of three-needle clusters. Plekhanov therefore decided to look for a similar effect at Tunguska – and he discovered it! Pines with three needles in a cluster did occur more often near the Southern swamp, their number diminishing with distance from the epicenter of the Tunguska explosion. And the maximum number of pines with three needles in a cluster was found to be where there was the maximum amount of ytterbium on Ostraya Mountain. Also the second maximum was on the canyon where the Churgim Creek flows, where in 1927 Leonid Kulik had set up a camp of his first expedition to Tunguska. Subsequently, these findings have been corroborated by several expeditions organized by the ITEG, and a catalog of 5,000 entries of such pines has been compiled.

Does this mean that we are dealing here with some sort of mutation? Opponents of the nuclear hypothesis point to the fact that the same effect occurs after usual forest fires – which did happen at Tunguska. Generally, they are right. The percentage of three-needle clusters in pines may increase due to both causes – "ecological" (occurring after forest fires) and "mutational" (as, say, occurred in the zone of the Chernobyl disaster). Yet these causes can be reliably differentiated: the "mutational" effect is more intensive than the "ecological" one. At Tunguska its scale greatly exceeds usual "ecological" figures. For example, at the epicenter were found several pines with an unbelievably powerful anomaly: more than half of all clusters on these trees turned out to have three needles.

But the strongest evidence that the three-needle clusters in Tunguska pines are due to a genetic mutation is their inheritability. This effect does exist in pines that are the second and third generations of the trees grown in the taiga after the catastrophe. And it is only genetic mutations that may be inherited.

There also exists at Tunguska another genetic effect – discovered by Academician Victor Dragavtsev, who mathematically processed the data collected by ITEG scientists... Any living thing belonging to the same species has, naturally enough, like traits: say pine trees of the same age grow at comparable rates. Comparable, but not identical. In fact, these rates fluctuate around an average figure, these fluctuations depending first on individual hereditary characteristics of the trees and second on environmental conditions in which they are growing. In other words, the trait dispersion consists of two components: innate and acquired. To find out which of these two components we are dealing with is not that easy, but geneticists have developed mathematical methods that make it possible to discriminate between them. Early in the 1970s Victor Dragavtsev, then a scientific worker of the Institute of Cytology and Genetics in Novosibirsk and later the director of the largest genetic bank in the world – the N. I. Vavilov Institute of Plant-Growing in St. Petersburg – proposed a new mathematical method to perform this task.[24] Having taken an interest in the Tunguska problem he paid attention to the data accumulated in the above-mentioned catalog of 5,000 Tunguska pines. When examining the pines, the ITEG scientists measured 20 parameters of each tree, including their yearly growth rates. Dragavtsev decided to use these data for further processing with the help of his method.[25]

His main conclusion was that over an area of about 200 km^2 the frequency of genetic mutations increased by a factor of 12 over what is normal. The unknown agent promoting these mutations acted on this territory 10 times more effectively than gamma rays in control experiments. Again, the two peaks of the Dragavtsev effect fall on the Ostraya Mountain and the Churgim Canyon, just as for the three-needle clusters in Tunguska pines. In the mid-1990s, Dr. Yury Isakov confirmed Dragavtsev's result by a different method.

The answer to whether or not the Tunguska taiga trees underwent a genetic mutation could have been obtained rather simply. All one had to do was to analyze the DNA in the seeds of living pines. So, Academician Nikolay Vasilyev invited several researchers from the N. I. Vavilov Institute of General Genetics of the Russian Academy of Sciences to participate.

Dr. Olga Fedorenko carried out all necessary analyses and signed a research report, which stated that some genetic effects in the region of the Tunguska catastrophe have in fact occurred. To continue and develop this investigation, agreement was needed on the collaboration with Dr. Fedorenko's chief at the institute – Professor V. N. Shevchenko. Academician Vasilyev took the report to the professor and proposed that he carry out some joint research. Professor Shevchenko dismissed the matter with a wave of his hand: "Mutations at Tunguska? Absurd!" Then Vasilyev showed him the report signed by scientific workers of his own department.

The professor became somewhat nervous, being unable to explain anything, but remained adamant in his reluctance to conduct any genetic studies at Tunguska. When Vasilyev spoke to Dr. Fedorenko, she confirmed for him that both the initial data and her conclusions had been correct. As for the panic that had overwhelmed her chief, she did not understand its cause and definitely could not be responsible for it. But it seems that Professor Shevchenko was shocked by a scientific result that was both reliable and anomalous.[26]

It is appropriate, however, to ask one more question. Would the Tunguska mutations have occurred only in trees? What about the Tunguska fauna? True, animals in this region are few and far between, and those present at the time of the catastrophe have died – and their descendants could have left the area. But there are ants at Tunguska that lead, so to say, a very settled life. The ants living now in the region of the Tunguska explosion are, most probably, direct descendants of those living there in 1908. Having studied some characteristics of ants dwelling in various parts of the Great Hollow (the length and width of the head, the width of the eyes, and so on), geneticists V. K. Dmitrienko and O. P. Fedorova found that the insects living near Ostraya Mountain and at Churgim Creek did sharply differ from those caught in other places.[27] In other words, these differences were greatest where peaks of mutations in local pines were also greatest. This seems to be significant. It would therefore seem that the ancestors of these ants did also undergo mutations at the Tunguska catastrophe of 1908.

But again, this is not the whole story. Although this region of Siberia was then (and still is) very sparsely populated, it turned out that the Tunguska phenomenon affected human genes as well, not only those of trees and insects. In the 1960s, 1970s, and 1980s, the leading Soviet specialist in the field of human genetics, Professor

Yury Rychkov (1932–1998), carried out an ambitious program of composing the complete Atlas of Genetic Geography of the USSR. Rychkov had worked at the same N. I. Vavilov Institute of General Genetics, where subsequently the Tunguska findings of its own researchers were treated so badly. His expeditions, aimed at studying genetic pools of various peoples, traveled all over the country and one fine day came to the Evenks of central Siberia. And here, to the great surprise of Professor Rychkov, he met with a Rhesus-negative person.

Generally, Rhesus factor (or Rh-D antigen) is the name given to a special protein that is attached to the surfaces of red blood cells. Individuals either have it (85% of the population in Europe and North America are Rhesus-positive) or do not have it (15% are Rhesus-negative). It is dangerous for the fetus if it inherits from its father an Rh-D antigen that differs from that of its mother. Then its mother's organism mistakenly recognizes the fetus as something alien and begins to "fight" with it, which may lead to a miscarriage. This is the so-called Rhesus conflict.

Among the Mongoloid inhabitants of Siberia, Rhesus-negative persons are exceptions. But as it turned out, Olga Kaplina, then 47 years old, was Rhesus-negative, and her children died increasingly earlier with every childbirth – which is the typical pattern of a Rhesus conflict. Professor Rychkov had examined this case in detail and had come to the conclusion that the source of this conflict was a mutation that affected Olga Kaplina's parents, who had experienced the Tunguska catastrophe. In 1908 they lived between the Northern Chunya and Teterya rivers and were eyewitnesses to the event. Olga Kaplina gave her parents' impressions as "a very bright flash, a clap of thunder, a droning sound, and a burning wind."[28]

Nikolay Vasilyev (the leader of the Tunguska studies and a noted immunologist) thought that the conclusion of Professor Rychkov was probably correct. "Organisms and inhabitants of the territories that were several decades ago exposed to small dozes of ionizing radiation demonstrate similar genetic changes," wrote Vasilyev. "This occurs, in particular, in those areas of the Altai Mountains that experienced radioactive fallouts from the nuclear tests at Semipalatinsk."[29]

Thus, we can conclude that genetic mutations at Tunguska do exist – in trees, ants, and human beings – probably due to the

Tunguska explosion. There is, of course, more to do on mutations in the region; let's hope these investigations will progress. And since the ionizing (or hard) radiation is the most typical cause of such mutations, let's now return to the question of radioactivity at Tunguska, which was touched upon in Chapter 5.

One can frequently read or hear that this question was settled long ago: no increase of radioactivity in the region of the Tunguska explosion has been detected. In fact, this is not so simple. Just when, half a century ago, the search for radioactive isotopes at Tunguska commenced, the researchers expected to obtain an immediate and definite result: yes or no. But like the Tunguska problem in general, the problem of radioactivity turned out to be much more complicated and "shadowy" than had been imagined initially.

Dr. Alexey Zolotov, when starting his own studies of radioactivity at Tunguska, realized that measurements of radioactivity of the soils gave very uncertain results. He also understood that he would have to date exactly any discovered effect; otherwise it would be impossible to associate it with the Tunguska explosion. With this aim in view, Zolotov developed the method of layer-by-layer measuring of the radioactivity of tree rings. More than 1,000 samples of Tunguska trees were examined, and it was found that before 1908 there had been no traces of radionuclides. But immediately after 1908 there exists in tree rings a small but noticeable peak of radioactivity – produced, according to Zolotov, by the radioactive isotope Cesium-137, whose half-life period is 27 years. There is also a second peak – after 1945 – and this one is definitely due to American and Soviet nuclear tests in the atmosphere.

But how about the first peak? Could it be due to the Tunguska explosion? To agree with this conclusion would have been too risky. Critics assumed that radioactive fallout from the nuclear tests could have penetrated into the living trees and accumulated around the tree rings of 1908 that had been damaged by the blast wave. However, the peak of radioactivity dated 1908 has been found not only in living trees but also in those that had withered *before 1945*, when no contamination from atmospheric nuclear tests would have been possible.

Notice that the problem of Tunguska radioactivity was studied not by amateurs but by the most distinguished Russian radiochemists, in particular by Academician Boris Kurchatov, the father of

Soviet radiochemistry, and his close associate Dr. Vladimir Mekhedov. And they confirmed all results obtained by Zolotov.[30] First, the radiation effect did exist near the epicenter, but not far from it, being therefore a consequence of the Tunguska explosion. Second, the two peaks in tree rings proved to be real. And last but not least, the peak dated 1908 was found in the trees that by 1945 were already dead.[31] Alas, after the premature deaths of Academician Kurchatov and Dr. Mekhedov this line of research ceased.

In 1965, the famous American scientist Willard Libby, a Nobel Laureate and inventor of radiocarbon dating, attempted to verify the hypothesis of Lincoln La Paz, an American pioneer in the field of meteoritics, according to whom the TSB had consisted of antimatter.[32] Annihilation of such a body in the atmosphere would lead to forming a powerful neutron radiation that, in turn, would produce a considerable amount of radiocarbon ^{14}C. This radiocarbon would then be dispersed by air streams through the whole of the northern hemisphere. If the energy of the annihilation were about 25 Mt of TNT, the total amount of radiocarbon in the atmosphere would increase by 7%. And Libby did discover in tree rings of the years 1908 and 1909 (of two trees in the United States – one in Arizona and another in California) an increased concentration of radiocarbon.[33]

Some other scientists immediately tried to check the finding of such a world-renowned specialist, analyzing samples of wood taken in other places in the northern hemisphere. And they also succeeded. In particular, Libby's result was corroborated by Academician Alexander Vinogradov – an eminent Soviet geochemist and pupil of Academician Vladimir Vernadsky.[34] Increased concentrations of radiocarbon have been found in the Great Hollow as well. True, some authors associate it with a fluctuation of solar activity, not with the hypothetical ionizing radiation from the Tunguska explosion.[35]

Indeed, during a minimum of the 11-year solar activity cycle (i.e., a period when sunspots become rarer) concentrations of radiocarbon in the atmosphere usually increase. This is an empirical fact, although various astronomers explain it in different ways. And it so happened that such a minimum had fallen on the year 1909. However, the radiocarbon at Tunguska is distributed patchily, just as many other traces of this enigmatic event, which makes it difficult to explain in terms of the Sun's activity.

Besides, one should not forget about two important circumstances. First, the Tunguska explosion occurred at a considerable altitude – between 6 and 8 km over the ground. Judging from nuclear tests, radionuclides formed when atomic or thermonuclear charges detonate at such altitudes are swiftly dispersed in the atmosphere over the whole globe, only slightly contaminating the region of the explosion. Second, it had happened 100 years ago, and the first attempts to find radioactive traces were made half a century after the event when the sensitivity of the measuring equipment was rather low.

Equipment is now better, but the time interval from the moment of the explosion has obviously increased. Let's recall that just 10 years after the explosion of the American atomic bomb over Hiroshima (which was only about 13 kt of TNT but exploded only 580 m above the surface), there were no direct traces of radioactivity in the territory of the city. This is why American and Japanese physicists, who attempted in 1955 to reconstruct the picture of the radiation effects in Hiroshima, had to look for an indirect but more sensitive technique of measuring very weak radiation traces, which was called the method of thermoluminescence (TL).

By that time this was already used in geology for age determination of rocks and in archaeology for dating ancient ceramics. Some minerals, being exposed to hard radiation, store in their crystal lattice the energy of the radiation. When these minerals are gradually heated up to 400°C, they begin to glow, releasing the stored energy. This is the effect of TL. Analyzing the relationship between the temperature and the intensity of the emitted light (the TL pattern) one can obtain information about the geological history of the mineral. Naturally, while heated, the whole energy stored in the mineral is released, and therefore repeated attempts to heat it will not produce any TL effect. All information about its past is obliterated – and the mineral begins to accumulate new energy from radioactive sources surrounding it.

Archaeologists have excavated – and dated in this way – piles of ancient ceramic pots and their fragments. Ceramics are made from clay – and clay consists of minerals (in particular, feldspar), which is noted for its high thermoluminescent properties. While being produced, ceramic pots are subjected to annealing; consequently, the stored energy is wiped out by heat, and the material becomes "thermoluminescently blank." This moment is the starting point in its further "thermoluminescent history." Under the influence of various sources of radiation it begins to gather energy anew. The rate of this accumulation is known to specialists, so when an ancient ceramic pot is found its TL properties can be examined and its age determined. And vice versa. If the age of such an object is known, we can determine the dose of radiation that it has obtained during its history. In Hiroshima, this effect helped when examining the levels of TL of ceramic tiles from roofs to measure exactly the weak radiation effects around the epicenter of the atomic explosion.

Taking into consideration the very high sensitivity of this method, it was reasonable to use it at Tunguska for the same purpose. Nikolay Vasilyev hit upon this idea as far back as 1960, but it took a long time to put it into practice. Since the Evenk *chums* (Siberian tepees) were never covered by tile, and the Evenks themselves used pottery only rarely, the researchers had to concentrate their efforts on natural TL indicators – first of all, quartz and feldspars. These minerals, having wonderful thermoluminescent properties, are common in the Tunguska explosion area. If the explosion was accompanied by ionizing radiation, its TL influence can be traced.

Yet when trying to put this idea into practice, difficulties emerged. As distinct from ceramics, the thermoluminescent characteristics of natural minerals are very unstable. Radiation of dispersed radioactive elements, such as uranium, thorium, and radium, increases the energy accumulated in their crystal lattice, while the interior heat of our planet and the solar ultraviolet radiation release this energy and therefore reduce its amount. The resulting TL pattern is therefore far from unequivocal. And an additional flow of hard radiation (say, from a nuclear explosion) would have just changed a little this complicated picture.

Nevertheless, the ITEG member Boris Bidyukov, who had been running the research program "Thermolum" at the Independent Tunguska Exploration Group since 1976 (see his photo in Figure 8.5) and is still doing so, has cracked this problem. He designed and built four models of an installation to determine TL patterns of Tunguska rocks. On these installations Bidyukov has examined several hundreds of samples.

FIGURE 8.5. Boris Bidyukov, an engineer and psychologist from Novosibirsk, the long-standing head of thermoluminescent investigations at Tunguska that made it possible to discover traces of the hard radiation from the Tunguska explosion. Founder and chief editor of the *Tungussky Vestnik* (*Tunguska Herald*) journal (*Credit*: Boris Bidyukov, Novosibirsk, Russia.).

Thus it was discovered that within 10–15 km from the Tunguska epicenter the TL level considerably exceeded the background level. The zone of the increased TL level also has an axis of symmetry coinciding with the second TSB trajectory calculated by Wilhelm Fast. This trajectory runs almost directly from the east to the west. But apart from the abnormally increased thermoluminiscence, there exists within this zone a smaller area (some 5–6 km in radius) of a *decrease* in the TL level, as if superimposed on the former one. And the boundary of the zone of decreased TL coincides well with the boundary of the area of the thermal burn of the trees. Most probably, the decrease in the TL level was generated by the light flash of the explosion. (It heated the rocks and soils, reducing the thermoluminescent effect.)

But what about the *increased* TL? Is it just a fluctuation of the natural TL level or is it associated with the Tunguska explosion? Can we differentiate between these two possibilities? Yes, we can.

As was recently discovered, the artificially induced TL effect radically differs from the naturally induced one.[36] In nature, the energy inside the crystal lattice of minerals is accumulated gradually, but when there occurs a nuclear explosion its amount increases abruptly. If such a mineral is then exposed to a flow of ultraviolet radiation, the naturally induced TL effect is swiftly reduced, reaching the minimal level typical for this mineral. But the level of the artificially induced TL effect does not alter.

Boris Bidyukov has exposed to ultraviolet radiation a set of Tunguska samples, as well as a sample taken far away from the region of the catastrophe. He found that the ultraviolet radiation did not affect the TL of the Tunguska sample, as distinct from the TL of the control.[37] This means that the Tunguska explosion was probably accompanied by a burst of hard radiation.

Thus, the Tunguska event left behind, in addition to the flattened and burnt forest and geomagnetic disturbances, five smaller traces: the possible microscopic remnants of the TSB substance; anomalously fast post-catastrophic restoration of the taiga; genetic mutations in plants and other living things; radioactive fallout in tree rings; and evidence of the influence of hard radiation on local minerals and rocks. These are, however, no less important than the larger traces. These traces are material, objective, and reliable, and therefore they must be taken into consideration when creating models that are supposed to explain the nature of the Tunguska phenomenon.

Remember that the distribution of all these traces on the surface of the territory of the Great Hollow forms similar patterns around the epicenter of the explosion and the axes of symmetry of the leveled forest area. This regularity is further proof of their association with the Tunguska event.

But did these five smaller keys to the gate of the Tunguska fortress help us to get inside? Frankly speaking, more "no" than "yes." They have just limited the spectrum of possible interpretations of larger keys, restricting their "freedom of turning." If, for example, there are at Tunguska genetic mutations (a "small" key), then the "nuclear" explanation of such a "large" key as the local geomagnetic storm becomes more acceptable and its "ballistic" explanation less acceptable. When someone tries to turn a large or small key in the direction of a stony asteroid or a comet core, he or she can hear an unpleasant grinding. That's a wrong direction! On the other hand, the outlines of the strange space body (or bodies) flying slowly over the wastes of central Siberia in 1908 and exploding due to inner energy and emitting hard radiation become now somewhat more distinct. But the nature and mechanism of the Tunguska explosion still remain enigmatic.

Well, how then can we build the correct model of the phenomenon? The only way is by analyzing empirical facts and comparing them with the theoretical constructions developed by Tunguska researchers during the long history of this problem. But since the objective traces are not yet handing us a ready solution, it only remains to try and use the "subjective" information about the event. These are the testimonies of those people who saw the flight of a fiery body on the sunny morning of June 30, 1908, heard the sounds accompanying its motion through the atmosphere, and witnessed the final explosion. The amount of information is vast and instructive; perhaps it could help. Now let's proceed to its analysis.

Notes and References

1. Vasilyev, N. V. The Problem of the Tunguska meteorite on the verge of a new century. – *Transactions of the State Natural Reserve "Tungussky"*. Vol. 1. Tomsk: University Publishing House, 2003, p. 153 (in Russian).
2. Ibid., pp. 156–157.
3. See Bronshten, V. A. *The Tunguska Meteorite: History of Investigations*. Moscow: A. D. Selyanov, 2000, pp. 238–239 (in Russian).
4. See Kirichenko, L. V. About formation of the local fallout trail from the explosion of the space body in June 1908. – *Problems of Meteoritics*. Novosibirsk: Nauka, 1975, p. 121 (in Russian).
5. See Vasilyev, N. V. *The Tunguska Meteorite: A Space Phenomenon of the Summer of 1908*. Moscow: Russkaya Panorama, 2004, pp. 168 ff. (in Russian).
6. Golenetsky, S. P., Stepanok, V. V. Searching for the substance of the Tunguska space body. – *Interaction of Meteoritic Matter with the Earth*. Novosibirsk: Nauka, 1980 (in Russian).

7. Golenetsky, S. P., Stepanok, V. V. Comet substance on the Earth (some results of investigations of the Tunguska cosmochemical anomaly). – *Meteoritic and Meteor Studies*. Novosibirsk: Nauka, 1983 (in Russian).
8. See Kolesnikov, E. M. Isotopic and geochemical data prove that there is cosmic matter in the region of the Tunguska catastrophe. – *Tungussky Vestnik*, 2002, No. 15 (in Russian).
9. See Vasilyev, N. V. *The Tunguska Meteorite: A Space Phenomenon of the Summer of 1908*, p. 183.
10. Longo, G., Serra, R., Cecchini, S., Galli, M. Search for microremnants of the Tunguska cosmic body. – *Planetary and Space Science*, 1994, Vol. 42, No. 2.
11. See Vasilyev N. V., et al. *Noctilucent Clouds and Optical Anomalies Associated with the Tunguska Meteorite Fall.* Moscow: Nauka, 1965, p. 99 (in Russian).
12. See Vasilyev, N. V. *The Tunguska Meteorite: A Space Phenomenon of the Summer of 1908*, pp. 181ff; Boutron, C. Respective influence of global pollution and volcanic eruptions on the past variations of the trace metals content of antarctic snow since 1880's. – *Journal of Geophysical Research*, 1980, Vol. 85, No. C12.
13. Strictly speaking, yttrium is not a lanthanide, but their chemical properties are very close.
14. See Zhuravlev, V. K. Which elements did the Tunguska meteorite consist of? – *Tungussky Vestnik*, 1996, No. 1 (in Russian).
15. See Zhuravlev, V. K., et al. Results of heavy concentrate sampling and spectral analysis of soils from the area of the Tunguska meteorite fall. – *Problems of Meteoritics*. Tomsk: University Publishing House, 1976 (in Russian).
16. Zhuravlev, V. K., and Demin, D. V. About chemical composition of the Tunguska meteorite. – *Cosmic Matter on the Earth*. Novosibirsk: Nauka, 1976, p. 102 (in Russian).
17. Dozmorov, S. V. Some anomalies of the distribution of rare earth elements at the 1908 Tunguska explosion site. – *RIAP Bulletin*, 1999, Vol. 5, No. 1–2, p. 11.
18. Zhuravlev, V. K., and Agafonov, L. V. Mineralogical and geochemical examination of the samples of soils taken in the area of the Tunguska bolide's desintegration. – *The Tunguska Phenomenon: Multifariousness of the Problem*. Novosibirsk: Agros, 2008 (in Russian).
19. See Vasilyev, N. V., et al. Possible mechanism of stimulating the growth of plants in the area of the Tunguska meteorite fall. – *Interaction of Meteoritic Matter with the Earth*. Novosibirsk: Nauka, 1980 (in Russian).

20. Emelyanov, Y. M., et al. Utilizing multivariate analysis for assessing factors influencing the alteration of the rate of growth of trees in the area of the Tunguska meteorite fall. – *The Problem of the Tunguska Meteorite*. Vol. 2, Tomsk: University Publishing House, 1967, p. 136 (in Russian).
21. Golenetsky, S. P., and Stepanok, V. V. Comet substance on the Earth (some results of investigations of the Tunguska cosmochemical anomaly). – *Meteoritic and Meteor Studies*. Novosibirsk: Nauka, 1983, p. 118 (in Russian).
22. See Zolotov, A.V. *The Problem of the Tunguska Catastrophe of 1908*. Minsk: Nauka i Tekhnika, 1969, p. 131 (in Russian).
23. Golenetsky, S. P., and Stepanok, V. V. Comet substance on the Earth (some results of investigations of the Tunguska cosmochemical anomaly). – *Meteoritic and Meteor Studies*. Novosibirsk: Nauka, 1983, p. 118 (in Russian).
24. Dragavtsev, V. A., Nechiporenko, V. N. About distribution of genotypical deviations of statistically elementary features in plant populations. – *Genetika*, 1972, Vol. 8, No. 6 (in Russian).
25. This research work was conducted in the Laboratory of Population Genetics of the Institute of Cytology and Genetics. See Dragavtsev, V. A., Lavrova, L. A., Plekhanova, L. G. Ecological analysis of the linear increment of pines in the region of the Tunguska catastrophe of 1908. – *Problems of Meteoritics*. Novosibirsk: Nauka, 1975 (in Russian); Plekhanova, L. G., Dragavtsev, V. A., Plekhanov, G. F., Influence of some ecological factors on the manifestation of genetical consequences of the Tunguska catastrophe of 1908. – *Meteoritic Studies in Siberia*. Novosibirsk: Nauka, 1984 (in Russian).
26. See Vasilyev, N. V. Memorandum. – *Tungussky Vestnik*, 1999, No. 10 (in Russian).
27. Vasilyev, N. V., Dmitrienko, V. K., Fedorova, O. P. About biologic consequences of the Tunguska explosion. – *Interaction of Meteoritic Matter with the Earth*. Novosibirsk: Nauka, 1980, pp. 192–194 (in Russian).
28. See Rychkov, Y. G. A possible genetic trace of the Tunguska catastrophe of 1908? – *RIAP Bulletin*, 2000, Vol. 6, No. 1.
29. Vasilyev, N. V. *The Tunguska Meteorite: A Space Phenomenon of the Summer of 1908*, p. 206.
30. See Vasilyev, N. V., Andreev, G. V. Radioactivity at Tunguska. – *RIAP Bulletin*, 2000, Vol. 6, No. 2.
31. Mekhedov V.N. *On the Radioactivity of the Ash of Trees in the Region of the Tunguska Catastrophe*. Preprint 6-3311. Dubna: Joint Institute for Nuclear Research, 1967.

32. See La Paz, L. The Energy of the Podkamennaya Tunguska, Siberia, Meteoritic Fall. – *Popular Astronomy*, 1948, Vol. 56, pp. 330–331.
33. Cowan, C., Atlury, C. R., Libby, W. P. Possible antimatter content of the Tunguska meteor of 1908. – *Nature*, 1965, Vol. 206, No. 4987.
34. Vinogradov, A. P., et al. Concentration of ^{14}C in the atmosphere at the Tunguska catastrophe and antimatter. – *Reports of the USSR Academy of Sciences*, 1966, Vol. 168, No. 4 (in Russian).
35. See, for example: Nesvetaylo, V. D., Kovaliukh, N. N. Dynamics of concentration of radiocarbon in the annual rings of trees from the center of the Tunguska catastrophe. – *Meteoritic and Meteor Studies*. Novosibirsk: Nauka, 1983 (in Russian).
36. See Shlyukov A. I., Shakhovets S. A. On the validity of the TL-method of age determination. – *Tungussky Vestnik*, 1997, No. 7 (in Russian).
37. Bidyukov, B. F. The thermoluminescent imprint of the Tunguska event. – *RIAP Bulletin*, 1998, Vol. 4, No. 1–2; Bidykov, B. F. Thermoluminescent anomalies in the area of the Tunguska Phenomenon. – *Tungussky Vestnik*, 1997, No. 5 (in Russian).

9. Grasping the Chaos

Eyewitness testimonies occupy a very prominent place in the history of the Tunguska problem. To begin with they formed the basis of this subject. Without these testimonies, even if some were misquoted by newspaper reporters, Leonid Kulik would probably have never known about the Tunguska event, and it would have been forgotten forever. This could have happened if, for example, the Tunguska space body (TSB) had arrived from the north, where potential eyewitnesses were few and far between and nomadic Evenks in this wilderness had no contacts with newspapers. As for the 30 million leveled trees in the Great Hollow, they would simply have rotted – since nobody would ever have become interested in them.

Obviously, eyewitness testimonies are different from strict instrumental data. Useful information is but a "weak signal" hidden among different "background noises," and the researcher has to devote considerable effort to find this signal. What a pity that in 1908 there was no Prairie net or similar systems of automatic monitoring of bolide activity in the sky! But certainly, Tunguska eyewitness reports should not be ignored when looking for explanations of this phenomenon. As Dr. Vitaly Bronshten wisely noticed, we must reject even "good" theoretical models of the Tunguska event, if these models come into conflict with information obtained from the eyewitnesses. These reports can be considered as a kind of boundary conditions for the "Tunguska theories." If a theoretical model goes beyond these boundaries this means it has nothing to do with the real Tunguska phenomenon.

The researcher should, however, be careful. A judge in a court considering a criminal case does not wave away eyewitness testimonies, but neither are they accepted uncritically. Instead, he or she compares the different testimonies as well as the material traces of an event, filtering out possible eyewitness errors and spurious

V. Rubtsov, *The Tunguska Mystery*, Astronomers' Universe,

DOI 10.1007/978-0-387-76574-7_9, © Springer Science+Business Media, LLC 2009

information. And with time, the true picture of the crime may emerge. This is the path that is to be recommended for Tunguska researchers.

Eyewitness reports may be used not only as factual restrictions for Tunguska theories. They can also reveal such sides of the Tunguska event that have not been reflected in its material and instrumental traces – say, the outer appearance of the TSB. And only when all the three types of Tunguska evidence – material, instrumental, and informational – jointly corroborate a theory can the researcher be sure that he or she is building the correct picture of the phenomenon.

Now, what information do we have at present about observations of the flight and explosion of the TSB? Unfortunately, there are no longer any surviving eyewitnesses and therefore we are dealing only with written records of their testimonies. It is no longer possible to correct errors in these texts, nor to supplement them with any additional information. But the number of such accounts is large. The General Catalog of Tunguska eyewitness reports has 920 entries. It is based on materials that were published in newspapers, journals, and monographs, as well as on archival materials and first-hand information collected by members of the Independent Tunguska Exploration Group (ITEG), the Committee on Meteorites (KMET), and the All-Union Astronomical and Geodetical Society (AAGS) in their Siberian expeditions. When the catalog was being prepared for publication, 212 eyewitness reports were removed from it – reports that could not have had anything to do with the TSB. In all likelihood, the eyewitnesses saw other large bolides that flew over central Siberia in different years. But there remained 708 reports directly related to the Tunguska phenomenon. True, not every eyewitness account in the published catalog contains information about the *flight* of the TSB – in some reports, only sounds accompanying its flight are described, or the flash and the sound of the Tunguska explosion, or the post-catastrophic earthquake. Nonetheless, in about 500 accounts the witnesses report the flying body, describing its shape and/or its brightness and/or its direction of flight. Not all testimonies are sufficiently complete; alongside very detailed reports we can find those that say little more than "something did fly." But such accounts are also important. They mean that at the place where the witness resided, the TSB was in fact seen, which can help to determine its flight path.

Various accounts may also differ in their reliability and accuracy of the event. Having grouped reports in different categories and statistically analyzed them, the researcher may eliminate less-reliable and less-accurate accounts. But it would be a mistake to try and rank them in this way *before* analyzing them. Even the worst eyewitness has one essential advantage over the best investigator: he or she was there and the investigator was not. However, attempts to "correct eyewitnesses" were made more than once in the history of the Tunguska problem. Evgeny Krinov, as well as other KMET members, stated repeatedly that many witnesses of the Tunguska phenomenon "had muddled up the points of the horizon." This would have been strange for inhabitants of the taiga. Here is one example.

A. Bulaev of the Siberian city of Krasnoyarsk wrote in his letter to the USSR's Academy of Sciences, dated October 17, 1962: "In 1908 I lived, together with my parents, in the village of Verkhne-Pashinskoe, some 10 km from the town of Yeniseysk. On June 30, my aunt and I visited my grandma Marina who lived nearby. Two windows of her house faced south. While my aunt and grandma were talking, I was looking out of a window. Suddenly I saw a red ball with a fiery broom behind it. The ball was twice as large as the sun, and the broom emitted sparks. They were not that bright and swiftly dispersed in the air. I cried out: 'Look here! Little sun is falling!' All dashed to the window. The fiery ball was already going down behind the local graveyard and then both the ball and the broom vanished..."

Having thanked the eyewitness for the interesting information, the scientific worker of KMET, Igor Zotkin, nevertheless noted:

"We already know that the Tunguska meteorite fall was seen near the town of Yeniseysk. Your letter confirms this data. Indeed, at Yeniseysk and other settlements at the mouth of the Angara River the flight of the Tunguska bolide was observed by many people. Unfortunately, there are in your letter some errors as well. Probably, you saw the fiery ball in the east, not in the south..."[1]

Of course, during 54 years that passed between the Tunguska event and Bulaev's contact with the KMET people, the eyewitness could have forgotten which point of the horizon had faced the windows of his grandma's house. After examining all eyewitness reports that came from Yeniseysk, this could have become evident.

But if such an examination *starts* from correcting "a priori erroneous" information in these reports, how could we have hoped to obtain from them any objective data about the event? An attorney in a law court will do everything that can be done to make the judge believe in the version of the affair that is favorable for the person being defended, but of course a serious scientist cannot behave in a similar manner.

But all the same, the question of reliability of eyewitness testimonies does deserve attention, and we should consider it in some detail. These accounts were collected in three stages. First, immediately after the event: the questionnaires of Arkady Voznesensky and newspaper articles of July 1908. Then, 15–30 years later: interrogations of local inhabitants by Leonid Kulik, Evgeny Krinov, and Innokenty Suslov in the 1920s to the 1930s. And finally, 55–65 years after the Tunguska catastrophe: special expeditions of the ITEG, KMET, and AAGS. As regards their reliability and completeness, each of these sets of data has its own advantages and drawbacks.

Let's start from the first set of eyewitness accounts, collected in 1908. The Tunguska event had just happened, and therefore neither could it be forgotten nor could the TSB be mistaken for something else. If the gathering of data on the observations of the TSB flight had started immediately, the results obtained would have been comprehensive and precise. Alas, this did not happen, and therefore the information we possess is pretty muddled. Although Voznesensky's questionnaires contain very valuable material, his questions were aimed at getting information about an earthquake. Perhaps, because of that, among 61 answers only 11 mentioned the flight of the TSB. Newspaper articles of the time also deserve attention. Journalists happened to describe the Tunguska event in some detail. But they reported no individual eyewitnesses with their names and addresses. Instead, we find on these old yellowish pages mainly references to some unnamed persons. "Here people saw...," or even "They say that here people saw..." From the famous article by Alexander Adrianov, which had been published in the newspaper *Sibirskaya Zhizn* (*Siberian Life*) and subsequently drew Leonid Kulik's attention to this phenomenon, we can see to what extent this information could become corrupted. But certainly, not all reporters were prone to such fantasies, and even Adrianov himself had probably not invented the whole story. Perhaps it was told to him by a passenger from the train that had been stopped by its driver when approaching the station of Filimonovo who was frightened by the sounds of the bolide's flight.[2]

The second set of eyewitness reports was accumulated at a time not too distant from the event and more methodically. But the only thing Leonid Kulik longed to know was: where had the meteorite fallen? Its trajectory was for him of secondary importance. For him, a meteorite could only travel in one way – straight to the point where it was doomed to end its life. And being a very goal-oriented person, Kulik simply wished to find out where that point was, in order to dig up the meteorite. As for Evgeny Krinov, he just recorded for his future book *The Tunguska Meteorite* some stories told to him by people in Siberia. Krinov believed that to determine the trajectory of the TSB (from which it would become possible to calculate its orbit in the Solar System), several detailed eyewitness reports would be enough. So why would he have had to accumulate hundreds of such reports? For a "normal" meteorite, Krinov's approach would have been justified, but not for the TSB. The material collected in the 1920s and 1930s, although useful, was not systematic enough to definitely determine the TSB trajectory.

The third group of Tunguska observations emerged somewhat unexpectedly. By the early 1960s the Tunguska researchers considered the collection of new eyewitness reports as rather pointless. Most of the eyewitnesses had already died and those surviving would hardly remember anything useful. Such was the general opinion. The real situation was different. At that time in central Siberia there were still many people who had seen the Tunguska bolide and heard the terrible boom of its explosion. The whole event had been fixed firmly in their memories. This – no exaggeration – discovery was made by Victor Konenkin, a school teacher of physics from Vanavara, the settlement closest to the epicenter of the Tunguska explosion (see Figure 9.1). Konenkin was born and grew up in the village of Preobrazhenka, on the riverside of the Nizhnyaya (Lower) Tunguska River, where in the long winter evenings he heard so often the tales of his older neighbors about the striking event of half a century before.

In 1962, the teacher decided to find out what the enigmatic flying object had looked like and how it had flown. He traveled to dozens of villages on the Lower Tunguska and its tributaries,

FIGURE 9.1. Victor Konenkin, a schoolteacher from Vanavara who has discovered that the flying Tunguska space body had been seen not only to the south from the Great Hollow, but to the east as well, up to 500 km from this site (*Source*: Zhuravlev, V. K., Zigel, F. Y. *The Tunguska Miracle: History of Investigations of the Tunguska Meteorite*. Ekaterinburg: Basko, 1998, p. 124.).

interrogating the surviving eyewitnesses. If the eyewitnesses still lived at the same settlement where they had seen the TSB, Konenkin asked them to come to the place of their observation. They took with them a compass and an angle gauge. The eyewitnesses showed the teacher at which point in the heavenly sphere they had noticed the fiery body for the first time and where it had disappeared. Of course, some eyewitnesses had already forgotten details of their observations, but all of them remembered the flight of the fiery body and also whether it had flown from left to right or from right to left.

Konenkin's investigations enabled him to determine where the TSB had traversed the Lower Tunguska River. The task was accomplished very simply. This part of the river flows almost strictly from south to north, so that eyewitnesses located upstream (farther south) from the place where the TSB was traversing the river saw it flying from right to left, while those downstream (farther north from the intersection) saw the TSB flying from left to right. After processing the data collected, it turned out that the TSB had flown over the river near the village of Konenkin's Preobrazhenka. And its inhabitants did confirm this, saying that the fiery object had flown directly over their village in 1908.

So a simple method obtained a result that must be correct. But there appears a problem: the village Preobrazhenka is situated at a distance of 350 km from the Tunguska epicenter and *almost directly to the east*. Most previous eyewitness reports were gathered to the *south* of the epicenter – up to a distance of about 1,000 km. How, then, could the TSB have approached the Great Hollow simultaneously from the east and also from the south?

The information collected by Victor Konenkin was so startling that it needed verification. Several expeditions – sent by KMET, ITEG, and AAGS – left for the Lower Tunguska, and they confirmed that Konenkin's data were correct. They also gathered additional eyewitness reports themselves. To the 35 accounts collected by Konenkin, another 150 were added.

Later, Tunguska investigators spread their questioning activities farther east – up to the Lena River. This work lasted until 1972, when it became evident that the "ore" had been mined and no new eyewitnesses could be found. So during several years, about a thousand people who in 1908 had lived eastward from the epicenter of the Tunguska explosion were questioned. There are now available about 550 eyewitness reports from the eastern sector, some 400 of which contain descriptions of the flying TSB.

The third set of observational data proved to be *very* informative. Its number of reports is more than three-fourths of the total, and these accounts were collected very thoroughly. The expedition's researchers were repeatedly using compasses and angle gauges to obtain quantitative data about the TSB path. The only apparent disadvantage of this set of data is its late collecting. The eyewitnesses were interrogated more than half a century after the event, being, at the same time, well familiar with the layout of their landscape.

Incidentally, in 1999 Konenkin's calculations of the TSB trajectory were again checked by the experienced meteor specialist Dr. Vitaly Bronshten. And he confirmed once more that the results were definitely correct. It was over the village of Preobrazhenka – or maybe a couple of kilometers farther south – that the TSB had been moving to the place of its destruction.[3]

Thus, eyewitness reports from the first group (about a hundred accounts dated 1908) are very reliable, since they were fresh, but they contain few specific details. Reports from the second group, about 75 collected in the 1920s and 1930s, are also rather reliable, being relatively fresh. And they contain more details. As for the third group, amassed in the 1960s (550 accounts), these reports, although collected later, are richer in detail.

Now it became possible to form on the basis of this enormous amount of material an authentic picture of the Tunguska phenomenon in general and the TSB in particular. For this, the eyewitness reports had to be statistically analyzed and condensed. If, for example, 90% of eyewitnesses had said that the TSB had looked like a bright white ball flying from the south to the north, this would have meant that we have a reliable and coherent picture of the phenomenon. The remaining 10% of reports describing it differently could have been considered erroneous.

Alas, such an ideal scheme has remained a dream. First, eyewitness reports varied greatly in their contents and terminology, which made their direct comparison difficult. True, some details proved to be consistent. For example, not one of the eyewitnesses reported that the TSB had a dense smoky trail, so typical for iron meteorites. (Such a trail accompanied the fall of the Sikhote-Alin iron meteorite in 1947.) Therefore, the TSB could not be an iron meteorite. But the researchers already knew that, since no pieces of meteoritic iron had been found in the Great Hollow. Much more interesting was to find out what the TSB *could* have been. Or at least, how did it look and behave.

The ITEG founding fathers Victor Zhuravlev and Dmitry Demin, together with Alexey Dmitriev, embarked on a study of the full catalog of the Tunguska eyewitness reports. Dmitriev, being a scientific worker at the Institute of Geology and Geophysics in Novosibirsk, had been for a long time engaged in computer analysis of the descriptions of geological objects made by prospectors. He therefore suggested using the same methods for examining the Tunguska accounts. Each one was dissected according to a formal scheme, and characteristics of the Tunguska phenomenon (time and duration of observations, shape, and color of the flying body, its direction of flight, and so on) were extracted. The resulting set of formalized information was analyzed with the help of computer programs.[4]

Now, which results have been obtained?

There were three main areas of eyewitness reports (see Figure 9.2). First from the southern sector where the TSB had been seen by inhabitants of settlements situated on the banks of the Angara River, second from the eastern sector (the upper reaches of the Lower Tunguska and Lena rivers), and third from the central area surrounding the epicenter of the Tunguska explosion – up to about 100 km from it. The "southern" observations were mainly collected before World War II, the "eastern" ones in the 1960s, and the "central" observations both

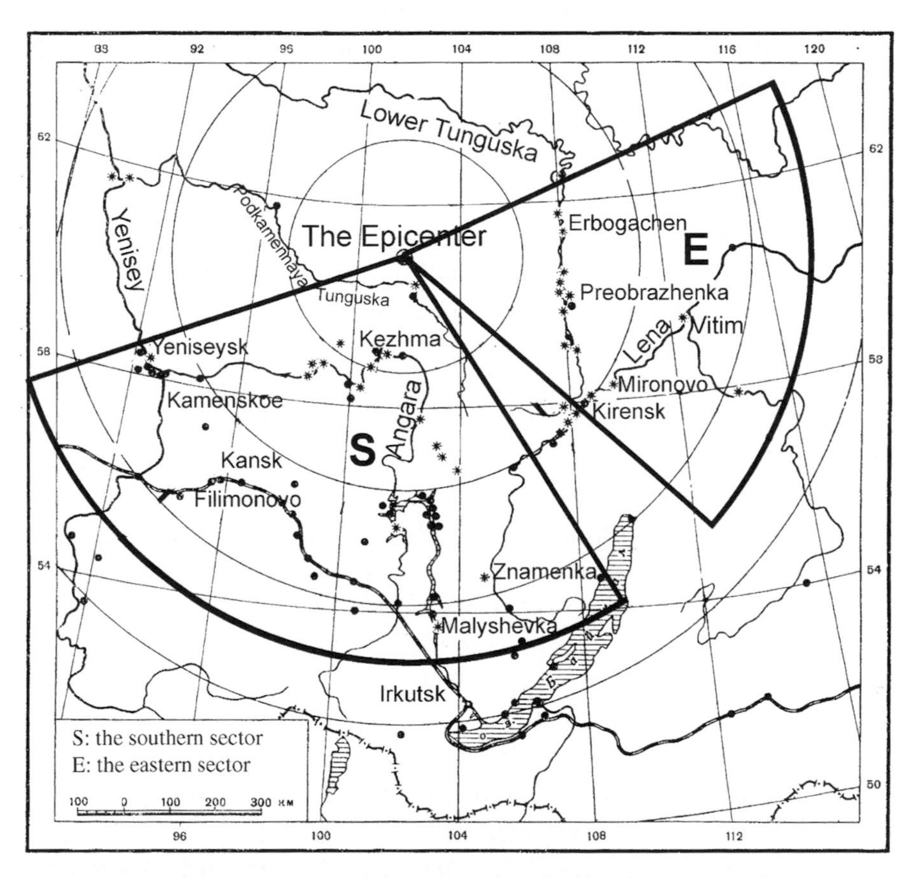

FIGURE 9.2. The southern and eastern sectors, from where came reports of eyewitnesses observing the flight and explosion of the Tunguska "meteorite" (*Based on*: Zotkin, I. T., Trajectory and orbit of the Tunguska meteorite. – *Meteoritika*, Vol. 27, 1966, p. 109.).

before and after World War II. Such a distribution of "observational zones" was understandable: just in the southern and eastern sectors the density of population in 1908 was relatively high, whereas to the north from the epicenter there were no permanent residents at all. But what seemed highly incomprehensible from this work was the radical difference between the images of the TSB built on the basis of the southern and eastern sets of observations. Data obtained inside each sector made it possible to create a statistically reliable and coherent image of the Tunguska phenomenon, but these two images were utterly different.

In the south, the phenomenon (including thunder-like sounds) lasted half an hour or more. The brightness of the TSB was comparable to the Sun. The body looked white or bluish. It had a short tail of the same color, and after its flight there remained in the sky iridescent bands resembling a rainbow and stretching along the trajectory of the body's motion. And it flew from the south to the north.

Take one example. In 1908, political exile T. N. Naumenko had lived in Kezhma some 215 km south-southwest from the epicenter of the explosion. In 1936, when in Moscow, he recalled: "The day was sunny and absolutely clear – not a cloud in the sky; no wind at all; complete silence. I was facing north. At about 8 o'clock the Sun was already quite high in the sky, when there was a hardly audible sound of thunder. It was far away but it increased. There was a weak clap of thunder and I quickly turned to the southeast, towards the Sun. Its rays were being crossed from the right by a broad fiery-white stripe. On the left an elongated cloudy mass was flying to the north. It was even brighter than the stripe – dimmer than the Sun's disk but almost as bright as its rays. A few seconds after the first clap of thunder, there was a second much louder clap. The flying lump was no longer visible, but its tail (the stripe) was now to the left of the Sun's rays. It was getting broader than it was when on the right. Almost immediately there followed a third clap of thunder, so powerful that the earth trembled and a deafening rumble resounded over the boundless Siberian taiga."[5]

Also in Kezhma, a local dweller, A. K. Briukhanov, did not see the flying body but noticed the iridescent trail behind it. "I was dressing after a bath and suddenly heard a loud noise. Half-dressed, I dashed to the street and immediately looked at the sky, since the noise was coming from above. And what I saw were blue, red, and orange bands running in the sky, as broad as the street. After some time the bands faded, the rumble rang out anew, and the earth quaked. Then the colored bands appeared again and again, after which they went to the north."[6]

But if we look at the eastern TSB observations, we find that the brightness of the flying body was much lower than the Sun (as eyewitnesses emphasized, "one could look at it while not blinking"). Its color was red and the shape was that of a ball or an "artillery shell" with a long tail. Usually eyewitnesses said simply: a "red fiery broom" was flying or a "red sheaf," and it was swiftly moving in the western direction, leaving no trace behind. The duration of this phenomenon (including the "firing" after its flyby) did not exceed a few minutes.

Here is a typical description of the TSB observation from the eastern group. In 1908 Feofan Farkov lived in the settlement of Erbogachen (330 km from the epicenter to the east-northeast, on the right bank of the Lower Tunguska River). "I heard a rumble and looked southward. There was flying in the sky a fiery sheaf. I noticed it when it was already to the southwest from Erbogachen. The fiery sheaf flew from left to right – that is, to the west. Although it was flying swiftly, I had time to make out that the body was elongated, its head darker, and behind the head there was a flame and then a bundle of sparks. After its flight, there remained in the sky no trace. Windows in Erbogachen were rattling. All the people were so frightened and they said: 'Armageddon has come!'"[7]

Now a normal bolide moving through the atmosphere is slowed down by the friction of the air and therefore its temperature and brightness are diminishing. Generally speaking, the TSB had to behave in the same way. That is, its brightness must have lessened and the color must have changed from white to red. In reality, eyewitnesses in Erbogachen (330 km from the epicenter) saw a *red* bolide, whereas those in Kezhma (215 km from the epicenter) saw a *white* one, which is the opposite to what would be expected. Well, perhaps the eyewitnesses might have perceived (or described) the outer appearance and even the color of the Tunguska bolide incorrectly, but at least they could tell us how the bolide moved. So the initial objective of those gathering Tunguska eyewitness reports was very simple. They wished to determine the direction of flight and the slope of the path of the Tunguska "meteorite." This would have made it possible to find its radiant on the heavenly sphere (that is, the point from which it came to Earth) and then its initial orbit around the Sun. Yet suddenly, the scientists met in this work with serious obstacles. Taken alone, the southern observations of the flying TSB were in good accordance, but the eastern group of eyewitness testimonies brought discord.

Initially, before the "eastern" testimonies came to light, the situation had looked more or less simple. The first attempt to determine the TSB trajectory was made, soon after the Tunguska event, by Dr. Arkady Voznesensky, Director of the Magnetographic and Meteorological Observatory at Irkutsk. Having processed the data he possessed, Voznesensky concluded that the Tunguska meteorite had flown practically from the south to the north, with a small deviation to the east. Subsequently it turned out that Voznesensky's trajectory, being drawn on a map, passed within 70 km of the true Tunguska epicenter – a reasonably good calculation, one must admit. Leonid Kulik, during his meteoritic expedition of 1921–1922, talked with a number of eyewitnesses and was also certain that the Tunguska meteorite must have flown from the south to the north. The noted meteor specialist Igor Astapovich in 1930–1932, during his geophysical expeditions to the Angara River, collected new eyewitness accounts. He afterward processed all materials that were known by that time and came to the same conclusion: the TSB trajectory practically ran from the south to the north, with a very small deviation to the east.

What is more, Astapovich found that in Malyshevka (located some 800 km to the south-southeast from the epicenter) the TSB had flown from right to left, whereas in Znamenka (140 km to the northeast from Malyshevka) it flew from left to right.[8] Consequently, the TSB trajectory must have passed between these settlements. That is, the TSB did come from the south and move almost precisely to the north. But at the same time, it must have passed, according to Konenkin's findings, over the village of Preobrazhenka, which was located 350 km from the epicenter almost directly to the east. That is, the TSB came from the east and moved almost precisely to the west.

Now, what did the Tunguska researchers achieve, having accumulated a whole lot of testimonies of eyewitnesses of the TSB flight in the southern and eastern sectors of the region, and having composed the complete catalog of these accounts and statistically processed the data? They obtained *two different TSBs*, one of which was relatively slow flying to the Great Hollow from the south, shining with bright white-bluish light, whereas the second one was racing from the east, glowing red. Not bad! Does this mean that two giant bolides flew on very different trajectories to the same final point on the same morning? Not probable, at least if we are dealing with natural bodies from space.

So how do we resolve the paradox that has come from the detailed study of the eyewitness reports?

Well, it seemed reasonable to consider one of the sets of eyewitness testimonies as having nothing to do with the real Tunguska phenomenon. Either the southern or the eastern reports would have had to have been erroneously associated with it. But which? The answer looked obvious. Of course, it was the eastern set of observations that had to be discarded. The southern set is basic. It was collected while the scent was hot – and it was due to these eyewitness reports that Leonid Kulik reached the place of the TSB explosion and found there the enormous area of radially leveled forest. To consider these observations as having no relation to the Tunguska event would have been absurd. Whereas the eastern set, though rich and very systematically accumulated, was gathered more than half a century after the Tunguska explosion. Were it not for Victor Konenkin, these reports would have vanished with time, together with the eyewitnesses, and hardly any researcher would have supposed that they had ever existed. The "eastern testimonies" are excessive; they make a mess of the Tunguska problem instead of helping to solve it. Therefore, it is these reports that should be dropped and forgotten. Let's suppose that they had been due to the flight of another large bolide sometime in the 1920s, 1930s, or 1940s.[9]

This solution might have been accepted by the Tunguska research community. They could even have ignored the fact that the eastern eyewitnesses all point to 1908 as the year of the event – not to any other year or decade. Human memory, you see. But this simple solution ran into a serious obstacle. The most reliable traces of the Tunguska phenomenon are material ones – the area of leveled forest, first of all. And we know that the second Fast's TSB trajectory, determined from the axis of symmetry of this area, does run from the east to the west. Also in the same direction runs the TSB trajectory determined from the axes of symmetry of the zones of light burn and the thermoluminescent anomaly. These facts do demonstrate that over the Great Hollow the TSB was flying from the east to the west. Consequently, it is the eastern set of eyewitness testimonies that definitely has direct relation to the Tunguska phenomenon.

But here we have a big problem because we have already made sure that the southern set is also directly related to the flight of the TSB. So what about our analysis? It appears that neither the southern nor the eastern set of eyewitness testimonies can be justifiably discarded, yet they each tell different stories. But then, perhaps the TSB made a maneuver? If its flight path was winding, this might explain the drastic contradiction.

The question about possible TSB maneuvers was raised by astronomer Felix Zigel in a paper read at the Sternberg State Astronomical Institute in 1967. By that time Zigel was already aware of Konenkin's findings. He understood that the TSB had flown over the Lower Tunguska River near the village of Preobrazhenka, which is almost directly *east* from the epicenter. But he also knew that the TSB was seen at the village of Kezhma, almost directly *south* from the epicenter. Zigel drew attention to an interesting detail: nobody had seen the flying TSB to the north from Kezhma. Perhaps, having flown over Kezhma, the TSB turned to the east and then to the northwest – moving, so to say, in a zigzag course? In this case, of course, it could not be a natural body; rather, this maneuver seemed to corroborate Kazantsev's starship hypothesis.

In principle, Zigel's idea was reasonable. One maneuvering TSB looked more acceptable than several flying from different directions to the same final point. But the lack of eyewitness reports about the TSB flight between Kezhma and the epicenter could be explained in a simpler way – too sparse a population. Second, no one saw the flying TSB between Kezhma and the Lower Tunguska River. And third (perhaps the most important), when speaking before the leading Soviet astronomers Zigel did not know that Preobrazhenka was not the farthest eastern point where the flying TSB had been observed.

It was in the summer of 1967 that the ITEG-9 expedition, led by Lilia Epiktetova, questioned inhabitants of several villages by the Lena River and discovered that the TSB had flown over this river near the village of Mironovo, at a distance of 500 km southeast of the epicenter. If Mironovo had been situated farther north from Kezhma, such a maneuver would have looked like a simple zigzag. But Mironovo is situated *farther south* from Kezhma – and therefore, to get there, the TSB would have had to perform a very complicated series of turns.

There was another convincing argument against any maneuver of the "southern" TSB: the precise determination of the direction to the epicenter of the Tunguska explosion, which was made by Arkady Voznesensky from answers to his questionnaires. Of all TSB trajectories calculated by various scientists, it is the trajectory proposed by Voznesensky that deserves our confidence. When calculating it, he did not know where the Tunguska meteorite had ended its flight path. All other researchers (Astapovich, Krinov, Konenkin, Epiktetova) proposed their trajectories when they were well aware of the final point of the trajectory – namely the Southern swamp. So their considerations somewhat resembled forcing the data to fit the known answer. As for Arkady Voznesensky in 1908, he did not know of the Southern swamp's existence, yet his calculated trajectory approached this swamp (and therefore, the Tunguska epicenter) to an accuracy of 70 km. But then, the "southern" TSB must have flown straight to the Southern swamp, not making any maneuvers.

At the same time, materials collected in the eastern sector appear to testify that the "eastern" space body did maneuver. Konenkin not only found that the TSB had flown above the Lower Tunguska River and Preobrazhenka but also determined that it had flown from the east-southeast to the west-northwest. But moving in that direction the TSB could not have arrived at the Southern swamp. Instead it would have missed by a hundred kilometers. Also, Mironovo, Preobrazhenka, and the epicenter do not lie along a straight line. To fly over these three points, the TSB must have traveled along a distinct arc.

Incidentally, there are five "eastern" reports in which eyewitnesses describe how the flying body changed its direction of flight. Here, for example, is the testimony of V. K. Penigin, who was born in 1893. His point of observation was the village of Kondrashino on the right bank of the Lena River (some 500 km from the epicenter to the east-southeast):

"Then I was a boy and helped to bring manure to the fields. We were upstream from the village. The fiery flying body was well seen. It resembled an airplane without wings, or a flying sheaf. It was as long as an airplane and flew as high, but more swiftly. The body was as red as fire or a tomato. It was flying horizontally, not descending, and passed in front of the cliff of Tsimbaly, at about two-thirds of its height. Then the body covered some 2 km more and made a sharp turn to the right, at a very acute angle."[10]

Possible explanations of such strange behavior of the TSB will be considered in the next chapter. Here we would only like to note that the simplest hypothesis – that this was an alien spaceship – is not the only acceptable answer. In fact, under certain conditions even an ordinary piece of stone from space could have changed its direction of flight.[11] Though here is another problem. The most distant point of observation of the TSB mentioned in the early eyewitness reports (that is, the most trustworthy reports) is the village of Malyshevka. It is located about 800 km from the epicenter to the south-southeast. It was just a few days after the Tunguska event that a member of Arkady Voznesensky's earthquake monitoring network informed him that the bolide had been seen there. Somewhat later (in 1921) Leonid Kulik found that the TSB had also been observed on the bank of the Yenisey River, some 960 km to the southwest from the epicenter.

Therefore, having entered Earth's atmosphere at a great distance from the point of its disintegration, the TSB covered about 1,000 km, flying, naturally enough, in a flat path. But all "ballistic models" of the Tunguska event require a steep trajectory near the epicenter.

How can we resolve this contradiction? Dr. Vitaly Bronshten assumed that the slope of the TSB path varied. For the most part the TSB was moving at an acute angle to Earth's surface, but near the Great Hollow, at an altitude of 30 km, it made a sharp turn down, with the angle increasing approximately from 10° to 40°. This could have happened if, due to the burning of the TSB as it rushed through the atmosphere, its shape changed and it began to resemble a *Soyuz* or an *Apollo* space capsule turned upside down. Then the aerodynamic force would have acted downward. (Such a space capsule is shaped like a truncated cone with a convex base. When normally reentering the atmosphere, the base is beneath and the aerodynamic force acts upward, making it possible for the spacecraft to fly in a flat trajectory.) The idea was attractive, since it allowed reconciling the seemingly incompatible parameters of the TSB trajectory at the beginning and in the end of its flight.

This solution led, however, to another contradiction – this time with hypersonics and the laws of the strength of materials. As Dr. Andrey Zlobin, chief of a department of the Central Institute of Aircraft Engine-Building in Moscow, noted: the crucial factors in Bronshten's model were the strength of the TSB material and the g loading (that is, Earth's gravitational effect plus the forces of acceleration during this maneuver). For comparison, the Russian fighter aircraft Sukhoi Su-37, built from special alloys and composite materials and having superb strength characteristics, may sustain up to 10 g loading. But the icy core of a hypothetical Tunguska comet, with a mass of about one million tons and flying at a velocity of 30 km/s,[12] would have changed its trajectory at the cost of aerodynamic forces for about 30° – when descending from an altitude of 30 km to an altitude of 8 km. And it would have done this in a couple of seconds. In this case, the g loading would have exceeded the normal terrestrial gravitation by *several hundred* times. Would the comet core have sustained this? Definitely not. "If you do not agree with this conclusion," remarks Dr. Zlobin, "it means you have made the epochal discovery: that supersonic aircraft may be built from ice!"[13]

In other words, even if a fragile cometary core had reached the altitude of 30 km, its attempt to make a sharp turn down would have immediately destroyed it. Meanwhile, it is well known (and well substantiated) that the TSB exploded at an altitude of 6–8 km. One could add that such a maneuver would have been quite as dangerous for a stony meteorite. So stone also is not a good construction material for supersonic aircraft.

By the way, there is in the Tunguska reports a strange detail: the eyewitnesses constantly say that they heard the sounds first and only then they saw the flying body. "This peculiarity was noticed by many independent witnesses," wrote Evgeny Krinov.[14] For a meteorite, as well as for any other material object flying at a supersonic velocity, such a sequence of events is impossible. Nobody could have heard the sound of its coming before seeing the body itself, because the speed of light is far greater than the speed of sound. So, Krinov said in his book:

"The eyewitness made a mistake. It was the other way round: he saw the flying object and then heard the sound." Yet when the "heavenly boom" rang out, some eyewitnesses were in their houses, having no intention of leaving them. Had they not heard the strange "clap of thunder," they would have remained inside. Therefore, the time interval between the initial sound and the appearance of the fiery body was large enough for them to come out and see the flying object.

Whether or not it would be possible to explain this strange phenomenon by referring to the so-called electrophonic sounds is still not clear. Electrophonic sounds (hissing, crackling, whistling) can accompany the flight of some (though far from all) large bolides. Initially, this was noted in 1719 by the famous British astronomer Edmund Halley in accounts of eyewitnesses of a huge bolide that had flown over England. However, he could not accept the physical reality of such sounds and decided that this was a purely psychological effect. During the following 200 years this opinion dominated. Probably, the first scientist who dared to reject it was the astronomer and Tunguska investigator Professor Igor Astapovich in 1925. The very term "electrophonic sounds" was somewhat later coined by Professor Pyotr Dravert (1879–1945), living in Omsk and also studying the Tunguska problem. (By the way, Dravert was a descendant of an officer from the army of the Emperor Napoleon Bonaparte, who had been captured in Russia in 1812 and never returned to France. In 1921–1922 Pyotr Dravert took part, together with Leonid Kulik, in the first meteoritic expedition through European Russia and Siberia.)

The nature and origin of these sounds are still vague, but the most popular theory, developed by the Australian astronomer Colin Keay in 1980, holds that such bolides are generating radio waves of very low frequencies, which, in one way or another, can be perceived by some people as audible sounds. However, the mechanism of this means of perception remains enigmatic.

Naturally, since radio waves move at the speed of light, electrophonic sounds generated by bolides would move far faster than the bolides themselves. However, they cannot be very loud – nothing approaching the sound of thunder. Usually electrophonic sounds are very soft, being described by witnesses as hissing or humming. In the above-cited observation of T. Naumenko, the first "clap of thunder" definitely preceded the appearance of the fiery body.

Thunder, roar, cannonade, firing – these are the words that were used most frequently – in three quarters of all accounts – by Tunguska witnesses describing sounds accompanying the flight of the TSB.

Perhaps this is how they perceived the ballistic shock wave produced by the TSB flying at a supersonic speed. It was strong to generate acoustic waves powerful enough to frighten people and even to perturb the water in the Angara River, but not so strong as to cause destruction. For example, in 1938 Leonid Kulik talked with D. F. Briukhanov who, in 1908, had lived not far from Kezhma. "I was plowing a field, recalled Briukhanov, and had just sat down near my wooden plow to have breakfast when heavy blows occurred – like the firing of pieces of ordnance. My horse fell on his knees. Above the forest in the north appeared a flame. I thought that some enemies were shooting... Then I saw firs bend down and decided that a hurricane had started. So I grasped my wooden plow with both hands not to let it be carried away. The wind was so strong that it blew soil from the field. And then this hurricane drove a large wave on the Angara. I saw all this very well since my field was on a hill."[15]

Of course, the ballistic shock wave could not have preceded the approach of the bolide itself. But neither have thunder-like electrophonic sounds been reported before. It is no mere chance that the catalog of electrophonic bolides that were observed over our planet between 1683 and 1984, compiled by Dr. Vitaly Bronshten and two colleagues, does not contain the Tunguska meteorite entry.[16] Dr. Bronshten, being a true specialist both in the Tunguska problem and in the problem of the electrophonic sounds, understood perfectly that the electrophonic explanation of the Tunguska thundery sounds was not tenable. So this enigma remains unsolved.

Needless to say, impressions of those eyewitnesses who were in the central area of the Tunguska explosion differed considerably from the impressions of distant eyewitnesses. The Evenks who were then still sleeping in their *chums* could not see the approaching space body, but they heard in a doze the noise accompanying its coming, to be awakened by the Tunguska explosion itself – or even, according to some eyewitness accounts, by a series of explosions. And not only was it the boom that awoke them but also the blast wave that brought the *chums* down and threw them up into the air, scattering their suede covers and stunning their inhabitants.

In particular, the brothers Chuchancha and Chekaren, being young and healthy men, having crawled out from under the remains of their *chum* and standing on the bank of the Avarkitta River (some 30 km from the epicenter), swiftly gathered their wits and began to look around. They remembered the sequence of events very well. That morning they were woken by a few tremors, whistling, and a loud sound of the wind. Having gotten out from their sleeping bags, the brothers heard a "very great clap of thunder" and saw trees falling, their pine needles burning. After this they felt three more powerful bursts accompanied by bright flashes in the sky, and then a fifth burst at a great distance, farther north.[17] A fairly detailed and dispassionate description of a terrible event testified that Chuchancha and Chekaren had maintained their self-possession.

But older Evenks were simply stupefied and bewildered and did not realize what was happening. For example, the *chum* of the Evenks Ivan and Akulina stood at the mouth of the Diliushma River, some 35 km from the epicenter. Akulina told the ethnographer Innokenty Suslov about her experience in the following words:

> We were three in our *chum* – I with my husband Ivan, and the old man named Vasily, son of Okhchen. Suddenly, somebody pushed our *chum* violently. I was frightened, gave a cry, woke Ivan, and we began to get out of our sleeping-bag. Now we saw Vasily getting out as well. Hardly had I and Ivan got out and stood up when somebody pushed violently our *chum* once again, and we fell to the ground. Old Vasily dropped on us as well, as if somebody had flung him. There was a noise all around us, somebody thundered and banged at the *elliun* (the skins covering a *chum*). Suddenly it became very light, a bright sun shone at us, a strong wind blew at us. Then it was as if somebody was shooting, like the ice breaks in the winter on the Katanga River, and immediately after that the *Uchir* dancer swooped down, seized the *elliun*, turned it, twirled it, and carried it off – somewhere. Only the *diukcha* (the *chum's* framework, consisting of 30 poles) has remained at its place. I was frightened to death and became *bucho* (lost consciousness)...[18]

When she regained consciousness Akulina did not recognize her surroundings. Some trees lay on the ground; others stood without branches or without leaves. A box with plates and dishes was lying at a distance. It was open, and many cups had been broken. Fox pelts, squirrel skins, and ermine were hanging scorched on the twigs of larches. Dry trunks, branches, and deer moss were burning on the ground. Akulina's husband Ivan was wounded: he had been blasted away from the *chum* for about 40 meters and his arm was broken. Akulina and the men moved toward another *chum* of theirs by the Dilyushma River. But both this *chum* and a *labaz*, in which food and fishing nets had been stored, had also been destroyed by the fire, and they had to move on toward the Chamba River.

"When we reached it," she said, "we were already very weak. And we saw around us a miracle, a terrible miracle. The forest was not our forest. I have never seen such a forest in my life. It was so unfamiliar. We had had here a dense forest, a dark forest, an old forest. And now there was in many places no forest at all. On the mountains all the trees were lying down and it was light; one could see far away. And it was impossible to go by the mountains through the bogs because some trees were standing there, others were down, still others were bent, and some trees had fallen one upon another. Many trees were burnt and smoking."[19]

It's hardly surprising that frightened people who were so close to the epicenter of a 50-Mt explosion first of all tried to escape and paid little attention to what was happening in the sky. Rather, it is remarkable that not counting the deaths of the many deer belonging to the Evenks, there were no human casualties during the Tunguska catastrophe (apart from the old Evenk Lurbuman, who after the explosion sent his son Ulkigo to find out what had happened, and having heard his report about the huge scale of devastation "became scared to death and died").[20]

Nevertheless, one of the "central" eyewitnesses did see the flying TSB, even though it was flying, so to say, in a wrong direction at a wrong time. Ivan Aksenov, an elderly Evenk man, was one of the people who were questioned by the teacher Victor Konenkin during his trips to the upper reaches of the Lower Tunguska River. Before 1917 he had been a Tungus shaman, a profession strongly disapproved of by the Soviet regime. So, after the Revolution he had to hide for many years in the taiga. Even many years later he had little liking for anything official – even for meteoritic expeditions sent from Moscow. But toward Konenkin the former shaman felt trust – perhaps because the teacher himself was half-Evenk – so he told him about his experience.

In 1908 Aksenov was 24 years old. That June morning, he was hunting near the mouth of a tributary of the Chamba River, some 25 km to the south-southeast from the catastrophe epicenter. Having shot an elk, he began to flay the carcass when suddenly all around "became red." Aksenov took fright, threw up his head – and at this moment "there was a blow." For some time he lost consciousness. "As I came to myself," recalled Aksenov, "I saw everything was falling around me, burning. I am lifting up my head and see devil's flying. The devil itself was like a billet, light color, two eyes in front, fire behind. I was frightened and I prayed, not to the heathen god but to Jesus Christ and the Virgin Mary. After some time praying I recovered: everything was clear. I went back to the mouth of the Yakukta where the nomad camp was. It was in the afternoon that I came there..."[21] The "devil," according to the old shaman, was going faster than airplanes now do. While flying it was saying "troo-troo" (not loud) and its direction of flight was down the Chamba, that is, north to south.

Whether or not Aksenov's story deserves to be taken seriously is a disputable question. On the one hand, both his observation of a flying body *after* the Tunguska explosion and the reported direction of its flight – from the north to the south – provoke natural doubts. But on the other hand, when rejecting what seems to be impossible, the researcher takes a risk of throwing out the baby with the bathwater. Statistical analysis of eyewitness reports is certainly a good and necessary thing, but information obtained from the sole eyewitness who was lucky enough to find himself in the right place at the right time can outweigh a number of reports from less well-situated witnesses. And taking into consideration that the Tunguska catastrophe could have involved more than one body, we could probably accept Ivan Aksenov's story with some degree of trust, if not with unqualified reliance. As for the direction of flight of Aksenov's "devil," we can safely suppose that having survived the Tunguska explosion the eyewitness confused the points of the horizon, and the body did in fact fly from the east to the west. "Down the Chamba" is not that precise, since the river meanders.

Incidentally, even though it is usually thought that there were no eyewitnesses north from the epicenter of the Tunguska explosion, this is not so. As a matter of fact there was one witness who lived far from the Great Hollow. And although he did not see anything, he did hear something. This was ascertained by Ivan Suvorov – a Russian folklorist and writer, who from 1934 to 1965 led a nomad's life in Evenkya and Taymyr Peninsula, recording and translating into Russian legends of northern peoples. In May 1941, when in the upper reaches of the Khatanga River (which flows into the Laptev Sea of the Arctic Ocean) he met Christopher Chardu, a Yakut who in 1908 lived at the trading station of Essey, a distance of 850 km from Tunguska and directly to the north.

Chardu described to Suvorov his impressions from June 30. "The morning was very sunny. We were still sleeping. Suddenly some distant rumble rang out – again and again... And the wind sprang up over the tundra. I awoke and thrust out my head from under the blanket. Now I see that someone is lifting the *chum*. Not once but many times. So I swiftly ran out from the *chum*. There was nobody outside, but the wind was bending bushes to the ground... I was frightened and wondered what could it mean? Probably, Domogor [the heavenly tsar] was furious..."[22]

Of course, if the Tunguska phenomenon was confined to the arrival of a large meteorite and its explosion over the Southern swamp, there could have been no "distant rumbles" and "wind" 850 km to the north from the epicenter.

In general, the eyewitness accounts convincingly demonstrate that the details of catastrophe at Tunguska were more intricate than is usually supposed. In this respect, they supplement well the sets of material traces of this event – both "large" and "small." When analyzing these traces, researchers also begin to realize that past scenarios have proved unable to explain all the data. When we process the eyewitness reports, we obtain, instead of an unambiguous picture of a space body arriving from a definite direction, either two bodies flying in different trajectories or one body performing various maneuvers – or a combination of these.

Krinov's references to the "low reliability" of eyewitness reports and the inability of chance observers to determine even the main points of the horizon, to say nothing about the direction of flight of a bolide, do not sound convincing. Say the trajectory of the Sikhote-Alin meteorite was determined from the eyewitness accounts quite unambiguously, and no "dissimilar images" arose.[23] Yes, in this case the testimonies were collected soon after the meteorite fall and very systematically – but the scales of the Sikhote-Alin and Tunguska phenomena did also differ radically. Inhabitants of the Tunguska region remembered well the TSB flight and explosion even tens of years after the event, so very impressive had it been.

Certainly, overestimating the significance of eyewitness reports would be as wrong as underestimating them. Albert Einstein liked to say that correct physical theories cannot be directly inferred from experience. Actually, scientists invent their basic principles in a purely intuitive way, and then logically deduce consequences that can be empirically verified. And only these consequences are checked against the empirical facts. Of course, by "empirical facts" the great physicist meant results of properly performed physical experiments. But if it is difficult to create a good theory starting from data obtained in a laboratory, it is still more difficult to do the same from information where the signal is hardly more intense than the background noise. "Deep intuition" of the researcher is for this process no less important than "strict logic." Thus, attempts to "invent" unconventional theoretical models of the Tunguska phenomenon are in themselves far from blameworthy; yet the scientist should constantly compare theoretical schemes he or she is building with the real knowledge of the circumstances and consequences of the Tunguska catastrophe. To what extent the "Tunguska theories" developed for the last 100 years correspond to this knowledge, we will see in the next chapter.

Notes and References

1. Vasilyev, N. V., Kovalevsky, A. F., Razin, S. A., Epiktetova, L. E. *Testimonies of Eyewitnesses of the Tunguska Meteorite Fall.* Tomsk: University Publishing House, Moscow: VINITI, 1981, p. 248 (in Russian).
2. See Vladimirov, E. I. Meteorites in the basin of the Yenisey River. – *Interaction of Meteoritic Matter with the Earth.* Novosibirsk: Nauka, 1980, p. 232 (in Russian).

3. Bronshten, V. A. Trajectory and orbit of the Tunguska meteorite revisited. – *Meteoritics and Planetary Science*, 1999, Vol. 34, Suppl., pp. A137–A143.
4. Demin, D. V., Dmitriev, A. N., Zhuravlev, V. K. Informational aspect of investigations of the Tunguska phenomenon of 1908. – *Meteoritic Studies in Siberia*. Novosibirsk: Nauka, 1984 (in Russian); Dmitriev, A. N., Zhuravlev, V. K. *The Tunguska Phenomenon of 1908 as a Kind of Cosmic Connection Between the Sun and the Earth.* Novosibirsk: IGIG SO AN SSSR, 1984 (in Russian).
5. Naumenko, T. N. An observation of the Tunguska meteorite flight. – *Meteoritika*, Vol. 2, 1941 (in Russian).
6. Vasilyev, N. V., Kovalevsky, A. F., Razin, S. A., Epiktetova, L. E. op cit., pp. 72–73.
7. Ibid., p. 180.
8. Astapovich, I. S. About a possible trajectory and orbit of the Tunguska comet. – *Physics of Comets and Meteors.* Kiev: Naukova Dumka, 1965, pp. 109–110 (in Russian).
9. See, for example: Plekhanov, G. F. *Reflection on the Nature of the Tunguska Meteorite.* Tomsk: University Publishing House, 2000, pp. 26–27 (in Russian).
10. Vasilyev, N. V., Kovalevsky, A. F., Razin, S. A., Epiktetova, L. E. op. cit., p. 224.
11. See Rubtsov, V. V. *On the Trajectory of the Tunguska Space Body.* Manuscript. Kharkov, 1972 (in Russian); Khokhriakov, V. A. On the interaction of space bodies with planetary atmospheres. – *Kosmicheskiye Issledovaniya*, 1977, Vol. 15, No. 2 (in Russian).
12. Bronshten, V. A. On some methods of calculation of the blast wave and ballistic shock wave of the Tunguska meteorite. – *Interaction of Meteoritic Matter with the Earth*. Novosibirsk: Nauka, 1980, p. 160 (in Russian).
13. Zlobin, A. E. It is modern opponents of pioneers of the Tunguska problem who are under a misapprehension. – *Tungussky Vestnik*, 1997, No. 8 (in Russian).
14. Krinov, E. L. *The Tunguska Meteorite.* Moscow: Academy of Sciences of the USSR, 1949, p. 54 (in Russian).
15. Vasilyev, N. V., Kovalevsky, A. F., Razin, S. A., Epiktetova, L. E. op. cit., pp. 78–79.
16. See Bronshten, V. A., Grebennikov, V. S., Rabunsky, D. D. Catalog of electrophonic bolides. – *Topical Problems of Siberian Meteoritics.* Novosibirsk: Nauka, 1988 (in Russian).
17. See Suslov, I. M. Questioning witnesses in 1926 about the Tunguska catastrophe. – *RIAP Bulletin*, 2006, Vol. 10, No. 2, p. 17.
18. Ibid.

19. Ibid, p. 18.
20. Ibid, p. 19.
21. Vasilyev, N. V., Kovalevsky, A. F., Razin, S. A., Epiktetova, L. E. op. cit., p. 106.
22. Ibid., pp. 262–263.
23. Divari, N. B. Determination of the trajectory of motion of the Sikhote-Alin meteorite from eyewitness testimonies. – *Astronomichesky Zhurnal*, 1948, Vo. 25, No. 1 (in Russian); Divari, N. B. Phenomena accompanying the meteorite shower and its atmospheric trajectory. – *The Sikhote-Alin Meteorite Shower*. Vol. 1. Moscow: Academy of Sciences of the USSR, 1959 (in Russian).

10. From Comet to Plasmoid to Mirror Matter

The most general question about the nature of the Tunguska event may be stated very simply – what was it? Unfortunately, there is no simple answer. First of all, one can ask: Was it a cosmic phenomenon indeed? Since no one saw the Tunguska space body (TSB) outside the atmosphere, the very term "Tunguska meteorite" is just a metaphor. So we have the hypothesis that the TSB was enormous ball lightning – formally not an absurd idea, but after a closer inspection erroneous. Ball lightning remains a scientific mystery, and to explain the Tunguska enigma by another enigmatic phenomenon is not to explain it at all. Besides, no one has ever recorded any manifestations of ball lightning that would even remotely have resembled the Tunguska event. So this hypothesis is not realistic. Still farther from reality are such terrestrial models of the Tunguska event as the explosion of marsh gas, the eruption of a volcano, a somewhat unusual earthquake, and so forth. The only contribution these models make is a negative one. Their advocates have meticulously and persistently picked holes in other theories, which was definitely of some use for the development of normal reasonable models of the Tunguska catastrophe.

But how many hypotheses have been offered to explain this event? To determine their exact number would hardly be possible, since even serious scientists, who could be brilliant specialists in their own fields of investigation, occasionally attempted to solve the "so-called Tunguska enigma" after reading a couple of newspaper articles on the subject – and putting forward their own solutions in the same newspapers. Probably, the whole number of Tunguska hypotheses reaches a hundred, or so. But only about a quarter of them may be called scientific hypotheses in the strict sense of this word – that is, built according to the standards of science and with due consideration of empirical data. Not so few, after all, especially as these 20–25 hypotheses, being, as a rule, mutually inconsistent, have had to explain the same set of empirical data.

V. Rubtsov, *The Tunguska Mystery*, Astronomers' Universe,
DOI 10.1007/978-0-387-76574-7_10,

Also, the phrase "hypothesis to explain the Tunguska phenomenon" is somewhat vague. Researchers may agree between themselves about the nature of the TSB but disagree about the mechanism of its explosion. For example, a comet core could have entered Earth's atmosphere at a great speed and destroyed the taiga in the Great Hollow by its ballistic shock wave, but the core could also have exploded in the final stage of its flight due to thermal or chemical processes inside it. In this case the forest destruction would have been the result of joint action of both the ballistic and blast waves. So, even the hypothesis of the cometary nature of the TSB is in fact an array of several hypotheses. Nevertheless, if we pay attention only to the body's nature, temporarily setting aside the question of the mechanism of its explosion, the whole set of Tunguska hypotheses that have been put forward by now may be divided into the following three groups:

1. The TSB was one of the minor space bodies existing in the Solar System and known to astronomers (a meteorite or the core of a comet).
2. It was a *hypothetical* minor space body still not observed by astronomers, but probably existing in the Solar System or sometimes arriving here from interstellar space (a dense cloud of cosmic dust; a lump of "space snow" of extremely low density; a microscopic black hole; a "solar plasmoid"; an asteroid consisting of "mirror matter").
3. The TSB was an alien spacecraft.

And if we consider hypotheses about the mechanism of the TSB's explosion (or rather, about the cause of the forest destruction – since some of the proposed mechanisms cannot be called "explosions" in the proper sense of this word), they can be grouped as follows:

(1) The impact of a huge crater-forming meteorite. (This hypothesis has been convincingly refuted, but it did exist and was for a long time considered by meteor specialists as the only correct one.)
(2) The ballistic shock wave of a swiftly moving cosmic body that sharply decelerated in the air over the Southern swamp and collapsed into or ricocheted from the dense layers of the atmosphere.

(3) A thermal explosion.
(4) An explosion produced by the inner energy of the TSB (chemical or nuclear).
(5) A powerful electric discharge between the TSB and Earth's surface.

The majority of Tunguska researchers usually divide the question "What was it?" into two subquestions: "What kind of body was it?" and "How did it level so many trees in the taiga?" Traditionally, it is the area of the leveled forest that is considered the most important trace of the Tunguska phenomenon, whereas other traces (even "large" ones, such as the light burn of the vegetation and the local geomagnetic storm) are ranked as "auxiliary" traces. This is generally understandable: the area of the leveled forest was the first discovered trace of the event. It was found by Leonid Kulik in the 1920s and remained relatively unchanged until the epoch of the Independent Tunguska Exploratory Group (ITEG). At the same time, manifestations of the light burn of the vegetation that also impressed Kulik had disappeared almost completely by the late 1950s. As for the local geomagnetic storm, it was discovered somewhat "too late" to be considered as a trace of prime importance. So the great necessity of explaining the leveled forest and the lesser importance of explaining the light burn and the geomagnetic storm are psychologically understandable. Still easier to ignore, when developing Tunguska hypotheses, are "minor" traces of the Tunguska phenomenon such as the superfast restoration of the forest in the Great Hollow, anomalies of thermoluminescence, the paleomagnetic anomaly, and so on.

Incidentally, people trying to solve the enigma of the Tunguska "meteorite" have frequently forgotten that their "solutions" were nothing but conjectures. That's why there had appeared such funny newspaper headlines as "The Enigma of the Tunguska Meteorite Has Been Solved!" But to express even a plausible assumption about the nature of the Tunguska phenomenon is not the same as solving this enigma. Of course, any hypothesis must explain the facts associated with the Tunguska phenomenon, but what is definitely necessary is that the hypothesis is testable. And the best possible test for any hypothesis is its ability to predict some new empirical facts following from it *and not following from other Tunguska hypotheses*.

For such a big problem as the Tunguska event, a couple of successful predictions will hardly be sufficient, though. The history of the Tunguska problem has demonstrated this very convincingly. For example, Alexander Kazantsev predicted several facts that were in the 1940s regarded by meteor specialists as impossible and even absurd: the overground explosion of the TSB and the lack of any crater and meteoritic substance in the Great Hollow. These specialists believed that the Tunguska phenomenon was due to the impact of an ordinary crater-forming meteorite. According to their viewpoint, the TSB should have exploded when striking Earth's surface, forming a crater and leaving behind the remains of the meteorite. In fact there was neither – and the explosion itself did occur in the air. Thereby, the hypothesis of the crater-forming meteorite has been convincingly rejected, but the "spacecraft hypothesis" has not been proved. Why? Just because these three features (the overground explosion and the lack both of the crater and of the meteoritic substance) were for the spacecraft hypothesis necessary but not sufficient. They testified that the TSB was *not a crater-forming meteorite*, and only that. If, apart from crater-forming meteorites, only an alien spacecraft could have fallen to Earth, Kazantsev's hypothesis would have been proved. But this is evidently not the case. There exist in the cosmos other natural minor space bodies that could also collide with our planet.

Immediately, the meteor specialists rushed to create alternative hypotheses that could have explained the same facts, not going beyond the scope of the first group of suppositions – that is, that the TSB was another minor cosmic body existing in the Solar System and well known to astronomers. As we know, Fred Whipple's "dirty snowball" model of comet's core arrived in time. Such a body, generally speaking, could have exploded thermally or chemically, since it consisted of considerable amounts of watery ice and frozen gases.

With time it turned out that the "dirty snowball" had its own drawbacks in this respect. In the mid-1970s, Academician Georgy Petrov (1912–1987, one of the founding fathers of Soviet space technology, the creator of the thermal shield for Yury Gagarin's *Vostok* spacecraft, and the first director of the Institute of Space Studies) and Professor Vladimir Stulov at Moscow University repeatedly simulated the process of thermal explosions. They found that the icy core of a comet could not have exploded leaving no traces. By that time dozens of spacecraft had returned to Earth flying at great cosmic velocities, and several Soviet *Venera* space probes had landed on the surface of Venus, having penetrated through its dense atmosphere. Consequently, characteristic traits of the superfast atmospheric flight of material bodies were now understood much better than they had been in the early 1960s, when Professor Kirill Staniukovich and Dr. Valery Shalimov had devised their theory of the thermal explosion.

So what is needed for a flying body to explode in this way? In fact, only two things are necessary: heat must get to the body's interior at a faster rate than it leaves it and the flow of thermal energy must be powerful enough. Under these conditions a cosmic body will become overheated and explode while flying in the atmosphere and before hitting the ground. Ordinary iron meteorites, for example, are losing their speed and cooling down faster than they are heated, and therefore a thermal explosion is out of the question. In the so-called "zone of retardation" (at an altitude of about 15–20 km) their velocity is already practically zero, and they are simply falling down to Earth's surface under the influence of gravity. The Sikhote-Alin iron meteorite was unusually large and therefore it did not slow down but just broke into pieces due to the air resistance, and these pieces hit Earth at a sufficiently great speed to form dozens of craters. Petrov and Stulov's calculations show that only about 1% of the ballistic shock wave accompanying a cosmic body flying through the atmosphere is spent in heating its substance. Therefore no space body of normal density (even ice) could become overheated during its flight in the atmosphere. Rather, the 30 million Tunguska trees must have been leveled by the ballistic shock wave that separated from the TSB after it had collapsed due to the air resistance. And why not?

Just because the same calculations have demonstrated that the TSB could have completely collapsed and the ballistic shock wave could have done what it did only if the density of the TSB had been less than one-hundredth of the density of water. Such a body's mass must have been several hundred thousand tons, its diameter about 400 m, and the initial velocity some 40 km/s.

If Petrov and Stulov had been astronomers they would have realized they were wrong. Cosmic bodies with such a low density do not exist in the Solar System, or at least they are unknown to astronomers. The density of comet cores is, according to all existing models and observational data, about the density of water, which is a hundred times greater than the supposed density of the TSB. But Petrov and Stulov were specialists in celestial mechanics and hypersonics and did not worry too much about astronomical paradigms. Boldly they said: let's suppose that comet cores have just this super-low density. And they gave their model a name more beautiful than "dirty snowball." They called it the "cosmic snowflake." "Only this model," emphasized Petrov and Stulov, "could rationally explain all features of the Tunguska phenomenon."[1]

Astronomers were shocked. Objections rained down upon Academician Petrov and Professor Stulov. In particular, Staniukovich and Bronshten argued that even if, by a miracle, such a "cosmic snowflake" had originated in the Solar System, it would have been very quickly destroyed by the solar radiation, the solar wind, and the tidal effects of the Sun and large planets.[2] And in any case, it could not have passed several hundreds of kilometers through the terrestrial atmosphere and reached an altitude of less than 10 km. It would have dissipated much higher – at about 100 km above Earth.

The astronomers were definitely right: the very low mechanical strength of the "cosmic snowflake" would never have allowed it to reach the Great Hollow. And besides, astronomical data do rule out the possibility that comet cores could be low-density snowflakes. But Petrov and Stulov's main conclusion remains valid: a space body of normal density (consisting of ice or rock) would not have dispersed entirely in the air. Its fragments would have fallen onto Earth's surface, while a hypothetical body of super-low density would have dispersed completely at a great altitude – about 100 km. Now let's look at the real picture: there are no fragments of the TSB in the Great Hollow, but at the same time the TSB collapsed at an altitude not exceeding 8 km.

So what? This means, first of all, that the TSB was sufficiently dense and mechanically strong enough to fly through the whole terrestrial atmosphere. And second, the forest destruction in the taiga cannot be explained only by the action of a ballistic shock wave of a dissipating body. The body must have exploded and produced a blast wave as well – which is lacking in Petrov and Stulov's model of the Tunguska event. Their model is therefore incorrect.

But Petrov and Stulov have convincingly refuted the hypothesis of the thermal explosion of a comet core. Such a core would have left noticeable material traces on the ground. But these are not present. Should we therefore assume that the Tunguska explosion was nuclear, as Kazantsev and Zolotov had assumed? As a last resort, this hypothesis might just be considered.[3] But aren't there any "less exotic" options in the store of contemporary science and technology that could be used? Why not reanimate the old idea of Kirill Florensky's about the chemical explosion of a comet core – with due consideration given to the progress chemical explosives achieved during the years that have passed since this idea was suggested? In the 1970s the United States and the USSR developed effective new weapons – fuel–air explosives, also called high-impulse thermobaric weapons or vacuum bombs. Rumor is that the US military calls the vacuum bomb the Hellfire weapon, which is very apt because its explosive power fills the gap between nuclear and nonnuclear weapons.

How does this weapon work? Various industries have been damaged by vapor cloud explosion accidents, so military chemists hit upon the idea of using this principle for war. A bomb or warhead of a missile contains liquid fuel that is dispersed as an aerosol by the initial explosion. Then, this cloud of fine mist is ignited by additional charges, and the resultant fireball incinerates everything and everyone over an area of several hundred meters. The fireball heats the air to about 3,000°C, eating up the oxygen in the volume affected. When the hot gas rapidly cools the air pressure sharply drops, the inrush of air reaches great speed, and this destroys everything. Conditions necessary for the vapor cloud explosion are created with the help of special technical devices; but couldn't they occur naturally when the icy core of a comet was moving in the atmosphere?

Dr. Maxim Tsynbal, a chemist from Moscow, had a good understanding of such processes. Together with Dr. Vladimir Schnitke, a mathematician from St. Petersburg, he developed a model of the vapor cloud explosion of a comet core. In their theory, the core consisting of frozen gases (methane, acetylene, cyanogens, and others) is first broken up by the air resistance, forming a gaseous cloud that then detonates. This was not just another flimsy Tunguska quasi-hypothesis proposed for want of something to do and aiming rather at the self-advertisement of its authors than at solving the problem (such "hypotheses" have been legion). Tsynbal and Schnitke used figures and facts – first of all those having to do with the mechanical strength of the cometary substance. It is quite evident that the "cosmic snowflake" would not have reached the lower atmosphere, but what about a "normal" comet core whose density does not differ considerably from the density of water? Could it have reached the altitude of 6–8 km over Earth's surface, where the TSB exploded, moving at the velocity of several dozens of kilometers per second, a velocity needed for a "thermal explosion"? Petrov and Stulov did not consider this side of the question. They simply demonstrated that *if* the core had reached this altitude and exploded over the Southern swamp, then the mass of the substance falling on Earth would have been very considerable and easily detectable. But there is none, and therefore the core of the "Tunguska comet" had to have a very low density.

Tsynbal and Schnitke approached the problem from another direction, trying to find out if the icy comet core had any chance to penetrate the terrestrial atmosphere. The mechanical strength of ordinary watery ice is well known.[4] Calculations show that a monolithic icy body flying at a velocity of 10 km/s would have collapsed at an altitude of about 30 km. But a comet core could hardly be a monolithic body. In reality, its mechanical strength would have been much less and therefore it would have disintegrated higher.

Perhaps then, we should consider frozen gases in the composition of comet cores? Could they have helped the core of the "Tunguska comet" to overcome the air resistance and to reach its point of destruction? But the mechanical strength of frozen carbon dioxide exceeds by a factor of 2 that of watery ice. Even if we suppose that the core was monolithic and consisting entirely of this frozen gas, then moving at a speed of 10 km/s it would have collapsed at an altitude of some 20 km. And in any case, if the comet core did reach the point in which the TSB exploded it means that its speed did not exceed 2–3 km/s. Unexpectedly, Tsynbal and Schnitke confirmed Alexey Zolotov's conclusion that had been made on a very different factual basis – that is, from the structure of the area of the fallen trees. Their result did not even depend on the trajectory of the TSB flight through the atmosphere. No matter whether it moved in a flat or steep path, the speed of the comet core at the altitude of 6–8 km must have been low.

On the other hand, swiftly decelerating in the dense layers of the atmosphere, such a core could have evaporated almost completely and given rise to a vapor cloud. Its kinetic energy would have been quite sufficient for that. After this, the vapor cloud could have exploded due to a slight spark that could have originated in an electric discharge. Space bodies moving through Earth's atmosphere are electrified by friction with the atmosphere (just like amber – if rubbed with wool – but much stronger) and therefore such a spark could have occurred. If the mass of frozen gases in the comet core were, say, 10 million tons, this could have resulted in a 50-megaton explosion over the Southern swamp.

So, according to Tsynbal and Schnitke, the TSB did explode as a vacuum bomb, and its explosion leveled some 30 million trees in the taiga. As for the ballistic shock wave, it was weak. This is why the trees are lying strictly radially around the epicentral point. They have been leveled only by the blast wave, with no contribution from the ballistic shock wave (as had already been established by Alexey Zolotov). But how could the explosion of a moving vapor cloud having an enormous volume form a point-source epicenter? For a nuclear explosion – very short and having an enormous concentration of energy in the explosive substance – it would be possible, but hardly so for a vapor cloud one.

True, the hypothesis of Tsynbal and Schnitke does explain characteristics of the "second large trace" of the Tunguska phenomenon – that is, the thermal burn of leveled trees – better than the nuclear model. A nuclear explosion with a magnitude of 40–50 Mt of TNT occurring at a relatively low altitude would have been accompanied by such a powerful light flash that all the vegetation in the epicenter would most probably have been completely incinerated. In any case, the two larches that were found in 1958 at the edge of the Southern swamp and proved to be not only alive but devoid of burns never could have escaped such a fiery bath. But somehow they did escape it...

Temperature of the fiery ball of a thermonuclear explosion may reach, even if for a split second, some 10 million degrees. But the fiery ball of a vapor cloud explosion is much cooler – just about 3,000°C. It emits its energy mainly as infrared radiation, not as visible light. The infrared waves lie outside the visible spectrum at its red end, being sometimes called "black light" or "thermal rays." It was, therefore, the invisible infrared radiation that singed the vegetation in the Great Hollow and was perceived by local inhabitants as a skin burn.

Not all of the mass of the TSB would have evaporated at the moment of the main explosion over the Southern swamp, and the blast wave would have scattered its burning fragments throughout the Great Hollow. This would result in the patchiness of the after-catastrophe forest fire, well known to Tunguska researchers and putting them in a spot. Tsynbal and Schnitke have even proposed an acceptable explanation for the genetic mutations and anomalies of thermoluminescence discovered in this region. According to them, the chemical products of the Tunguska explosion rose to the ionosphere and, when going through the ozone layer, neutralized a large amount of ozone, forming there a "hole" open to radiation. Via this hole the high-energetic ultraviolet Sun radiation, usually absorbed by ozone, reached Earth's surface and affected the living organisms and local minerals.

Having studied Tunguska eyewitness accounts, the researchers concluded that the space body had flown from the south to the north, not from the east to the west. Somehow, the "eastern" set of eyewitness reports did not impress them. But yet, what about the axis of symmetry of the butterfly-like shape of the leveled forest area? It is a common opinion that this axis is the projection of the TSB trajectory at the final stage of its flight... And it goes from the east to the west, not from the south to the north. Here Tsynbal and Schnitke assumed that the flying TSB, heating due to the air resistance, evaporated very unevenly. Its shape changed, and the vapor jets, ejected from its surface, created a thrust. Consequently, the aerodynamic characteristics of the space body altered swiftly, and the body swerved unpredictably. So the body, flying generally from the south to the north, could have turned to the west when approaching the Great Hollow.

In fact, although the idea of "swerving" looks possible, the TSB could hardly have made such a complicated zigzag-like maneuver – turning after Kezhma to the southeast, then returning again on its path to the Great Hollow and overflying the Lena and Lower Tunguska rivers. The shape of the leveled forest area – the famous "butterflies" by Wilhelm Fast and John Anfinogenov – does not follow from this theory, either.

But its weakest point was that the comet core, which is surrounded by a cloud of gas and dust, the comet's coma, must have consisted of very pure ices – of water and frozen gases. Taking into consideration the enormous mass of the core (up to 10 million tons, according to Tsynbal and Schnitke), which had first to evaporate and then to explode, these ices had to be unnaturally pure. Supposing that the comet core had contained just 1% of silicate and metallic particles, there would have rested in Tunguska soil and peat 100,000 tons of hard cosmic substances. But as we know, in the Great Hollow lie just about *a ton* of such particles. Even if it is the real dispersed material of the TSB, and not simple fluctuations of the background fall of extraterrestrial matter, this figure utterly contradicts the hypothesis of Tsynbal and Schnitke. But recent astronomical investigations, supported by the data that were obtained by automatic space probes, do convincingly testify that the share of hard substances in cometary cores is fairly high – up to 50%. Therefore, the Great Hollow must have received up to 5 million tons of such substances. So where are they? Naturally enough, specialists in cometary astronomy have been very skeptical about the TSB model developed by Tsynbal and Schnitke.

Nonetheless, their work has contributed greatly to Tunguska studies. As a matter of fact, they have refuted the "classical" model of the thermal explosion of the swiftly moving comet core – demonstrating that under no conditions could such a core have reached the altitude of 6–8 km maintaining a high-enough velocity to have caused its thermal explosion.

In the 1980s the American astronomer Zdenek Sekanina also reasoned that the cometary hypothesis of the TSB is at variance with what we know about comets.[5] Being an astronomer and not a chemist, Sekanina did not attempt to develop a theory of the TSB vapor cloud explosion, but his calculations confirmed that the core of a comet would have disintegrated in the atmosphere at a much greater altitude than had in fact happened. Sekanina's additional argument against the TSB being a comet was its probable orbit in the Solar System. According to his calculations, it must have coincided fairly well with the orbits of asteroids from the so-called "Apollo group." As distinct from the majority of small bodies that revolve around the Sun between the orbits of Mars and Jupiter, these asteroids are moving in elongated orbits, traversing Earth's path in space. Not favoring the chemical model of the TSB explosion, Zdenek Sekanina came to a simple conclusion: since the TSB could not have been a comet core, it must have been a stony meteorite – that is, a fragment of an asteroid.

Well, let's suppose it was. How then could a piece of a stony asteroid have exploded in the air? As we know, a ballistic shock wave alone could not have leveled the Tunguska taiga, forming a butterfly-like figure. To form this the TSB must have exploded at the end of its journey through the atmosphere. However, it now seems that under certain specific conditions such an explosion of a stony cosmic body is possible – if, during its flight, it is swiftly fragmenting. A detailed theory of the fragmentation process was created in the late 1970s–early 1980s by Academician Samvel Grigoryan and Dr. Vitaly Bronshten.[6] Their theory was suitable both for a comet core and for an asteroid – substituting in the case of a comet the doubtful idea of the thermal explosion and in the case of an asteroid offering the mechanism for the explosion of an enormous stony meteorite.

Good. Now we have a theory explaining a very enigmatic aspect of the Tunguska explosion – the pattern of devastation it produced on the ground. Flying at a great velocity, a huge stony meteorite could have exploded in the air. And its strength characteristics would have allowed it, as distinct from a comet core, to reach an altitude of 6–8 km over the Great Hollow. But Vitaly Bronshten has asked the supporters of the stony meteorite hypothesis a simple question – where is the meteoritic substance? According to estimations, the overall mass of space dust at Tunguska does not exceed 1 ton at best. It is too little even for a comet core, but one could probably suppose that there exist in space comets with "very pure" icy cores. Although this hypothesis has not been proved as yet, it's not too fantastic. But for a stony meteorite with a mass of several hundred thousand tons at least, the lack of cosmic matter at the epicenter of the explosion seems inexplicable.

To put it bluntly, it is absurd. Had a stony meteorite exploded – due to thermal tensions or due to swift fragmentation – the Great Hollow would have been strewn with silicate dust, and the peat layer of 1908 would have contained lots of meteoritic matter. Not only dust, by the way. Calculations by Dr. Bronshten have proved that after such an explosion a great deal of large stony fragments – each weighing more than 10 kg – would have been present. These hard fragments would have fallen on Earth's surface for investigators to find and analyze. And where are these fragments at Tunguska? During recent decades very sophisticated techniques have been applied to find them, but this extensive search has gone unrewarded.

Bronshten was definitely right. It only remains to conclude with a touch of sadness that:

(1) The "snowflake" hypothesis by Petrov and Stulov does not work, first because there are no such "snowflakes" in the Solar System and also because such a snowflake would have disintegrated at an altitude of about 100 km, not at 6 km over Earth's surface.
(2) An icy comet core with a mass of one million tons could not have reached this point either – it would have broken apart at an altitude of about 25 km. To reach the necessary altitude at which the TSB exploded, this core would have needed a mass of 5 million tons at least. But in this case a question arises: Where is the "dusty" component of the cometary substance, whose share in comet cores, according to contemporary astronomical data, cannot be less than 50%?
(3) The same difficulty is met by Tsynbal's and Shcnitke's idea of the vapor cloud explosion of the evaporated comet core. Large amounts of the hard substance would have been present in the Great Hollow after this explosion and could have been easily found. Besides, Tsynbal and Schnitke believed that the main "explosive" in the Tunguska comet core was methane. But again, contemporary data indicate that there is only a small percentage of this gas in comets. For enough to be present to produce such a powerful explosion (with a magnitude of up to 50 Mt of TNT!), the overall mass of the comet core would have had to be several *dozen* million tons! And, again, where are millions of tons of cometary substance that would have fallen on the ground?
(4) And last but not least, the lack of appreciable quantities of silicate meteoritic matter in the Great Hollow does strongly contradict the hypothesis about a stony meteorite's explosion.

So, after decades of intense theoretical considerations and searches in the field, neither of the two main models of the

Tunguska phenomenon – cometary and asteroidal – can answer the most important (although far from the only essential) question: Where is the TSB substance?

What's to be done in this difficult situation?

Of course, coming to a temporary deadlock is not something unusual in studying complicated scientific problems, and the general strategy for such cases is obvious. We should look for new explanations of the phenomenon under investigation. But where and in which direction should these explanations be searched for? The spectrum of opportunities is rather broad; each researcher may find those fitting his or her own professional and personal inclinations. As often happens in science, the Tunguska investigators formed three different groups: conservatives, radicals, and anarchists. The conservatives paid their attention to the most obvious – and definitely important – question – whether or not all factors influencing a comet core or a stony meteorite flying through the atmosphere have been taken into proper account when analyzing this flight. Usually it was only aerodynamic forces that were considered – well studied and well described mathematically – but perhaps there is "something else" in the flight of meteorites?

The radicals behave more resolutely. If neither a comet core nor a stony meteorite can explain the Tunguska data, perhaps there exist in the Solar System (or sometimes flying through it) some cosmic bodies, still unknown to astronomers but having properties that could explain the Tunguska phenomenon?

And finally, the "anarchists." They lost heart and asked the ultimate question, that maybe there was no space body at all? Couldn't the Tunguska event be purely a terrestrial phenomenon? (The shade of Sergey Temnikov, who had participated in the Great Tunguska expedition of 1929–1930, after which he had sent a report to the authorities accusing Leonid Kulik of "inventing a fantastic meteorite" certainly went into raptures in this connection and agreed that there was no TSB.)

But let's look at the essential physical factors that could have been accidentally ignored by meteor specialists who were studying the flight and explosion of the TSB. One might have been the process of its electrification. Astronomers and meteor specialists did understand that this had to play *some* part in the interaction between the meteorite body and the air. It is thought, for example, that weak electrophonic sounds accompanying the flight of some large bolides could be explained by their electrification when traveling through the atmosphere. Perhaps this physical process could produce more powerful effects? Theoretically this is acceptable, but how can we measure the level of electrification of a piece of cosmic iron flying at a great altitude with an enormous velocity? We can't, so meteorite specialists have preferred not to include electrical effects in their theories and mathematical calculations.

The pioneer investigator of this question was Vladimir Solyanik – an engineer, not an astronomer. As far back as 1951 Solyanik read his paper at a meeting of the Commission on Comets and Meteors of the Astronomical Council of the USSR's Academy of Sciences, in which he tried to draw the scholarly community's attention to the missing factor of electrification.[7] Decades later, scientists became interested in his ideas, and he published his work in a collection of Tunguska papers.

Solyanik thought that iron meteorites could be shattered in the atmosphere not so much by the influence of the air resistance as by their electrification. They are too solid for aerodynamic forces to affect them. Say, for example, the Sikhote-Alin meteorite that fell in 1947 in the Soviet Far East had split during its flight into many large pieces, this disintegration starting at an altitude of 60 km. But the metal content of these pieces, which were collected by the expedition of the Committee on Meteorites, proved to be very strong and able to sustain much greater loads than the meteorite had been subjected to in the upper atmosphere. So why did it break up?

Solyanik pointed out an intriguing fact. When the Sikhote-Alin meteorite flew over a technician who was on a telephone pole repairing a telephone line he felt an electric shock. It seems, therefore, that the meteorite flying above the telephone line generated an electric field that induced an electric current in the line. Similar cases have been recorded when other large bolides have entered the atmosphere. If so, could such a field influence a meteorite itself?

Vladimir Solyanik has produced a simple but convincing theoretical description of the electric processes occurring when a piece of cosmic iron flies through the atmosphere. The molecules of air knock off electrons from the meteorite, which makes the meteorite lose its negative charge and acquire a positive charge. So the strength of the electric field around the moving meteorite swiftly increases to produce mechanical stresses in its substance. When the meteorite is approaching Earth's surface its positive charge induces a negative charge in the ground beneath the flying body, creating a zone of an electric field of increasing intensity. And as the electric charge of the flying meteorite rises, the altitude of its flight diminishes. Finally there arises between the meteorite and the ground something like a high-power electric ark, and the meteorite explodes.[8]

This is a good theoretical scheme – possibly fitting well some cases of bolide flights in the terrestrial atmosphere. But whether it has anything to do with the Tunguska phenomenon remains doubtful. First of all, according to Solyanik's computations, only an iron meteorite could acquire in its flight through the atmosphere an electric charge that would have produced such a powerful explosion. Stony meteorites could not do that – their physical makeup would not allow them to accumulate the necessary electrical charge. But had the TSB been an iron meteorite, the eyewitnesses would have seen a well-defined black tail consisting of small particles of meteoritic iron. Nobody reported seeing such a black tail. And once again, the same old question arises: Where is the meteoritic substance? Solyanik attempted to evade the issue by supposing that the TSB did not disintegrate completely over the Southern swamp, but that its main mass flew farther west and fell at a distance from the epicenter. This idea is interesting but hardly corresponds well with the enormous magnitude of the Tunguska explosion. Besides, *some* fragments of the iron meteorite would have been scattered near the epicenter as well, not only where the main mass of the TSB would have fallen. In 1951 one could assume that these fragments simply had not been found as yet; but since then this territory has been searched *very* thoroughly and no meteoritic iron has been found. Also the "electric explosion" would have lasted, according to Solyanik's calculations, not less than two seconds, while the TSB was still flying in a shallow trajectory with a great speed. But in this case, the leveled forest would not have been lain so radially. So, we must admit that Solyanik's electrical hypothesis (as well as its later variant developed by the rocket engineer Alexander Nevsky)[9] cannot explain even the most obvious empirical facts relating to the Tunguska catastrophe.

Then perhaps we should search for such an explanation in a more radical direction? Let's suppose that the Solar System contains "exotic" space bodies whose properties could help explain the Tunguska explosion? Generally speaking, even Petrov and Stulov's "cosmic snowflake" was an "exotic object" disguised as a comet core. That's why astronomers could not accept it as a possible solution of the Tunguska problem. Still more exotic is the "solar plasmoid" theory proposed by Alexey Dmitriev and Victor Zhuravlev. As Vitaly Bronshten noted with good reason, "If such bodies had existed, astronomers would have observed them. Diligent comet hunters would have discovered hundreds of such plasmoids per year. Nothing of this sort has ever happened."[10]

Well, it goes without saying that while both the "cosmic snowflake" and the "solar plasmoid" have been invented specially to explain the Tunguska phenomenon, they have never been seen and they lack interest for space scientists. But physicists have developed a lot of theories involving peculiar objects that may or may not exist in the cosmos. Take, for example, the ever-popular "black holes." In relativistic astrophysics, a black hole is a body (or rather a region of space) whose mass is so great that no material objects, not even photons, can escape its gravitational pull. Physicists showed that when a sufficiently massive star runs out of its nuclear fuel, it should collapse into a black hole. There is also observational evidence that some galaxies may contain gigantic black holes in their centers. Theoretically, as Stephen Hawking has calculated, there could also exist *microscopic* black holes that have survived from the early epoch of our universe.

So, in 1973, two scholars at the University of Texas in Austin, Albert Jackson and Michael Ryan, published in *Nature* a paper in which they suggested that the TSB might just have been one of these microscopic black holes – negligibly small but having a mass of one quadrillion (one followed by 15 zeros) tons. Such a super-dense body would have penetrated Earth and traveled right through, escaping from the Atlantic ocean somewhere in its northern part.[11] The idea got polite interest among physicists, who for some time discussed the question whether or not such microscopic black holes could exist. As for astronomers and specialists in the Tunguska problem, they did not take the idea seriously. If a small black hole made such a mess and leveled 30 million trees when entering Earth, then its exit from the ocean would have been accompanied by similar perturbations, including, most probably, a powerful tsunami that would have devastated the Atlantic coast of Canada and the United States. Happily enough, this did not take place, and no jumps of the atmospheric pressure had even been recorded. Thus, the hypothesis about the "Tunguska black hole" may serve as another good illustration of Vitaly Bronshten's words about abstract mathematical constructions – fairly scientific but having nothing to do with the Tunguska problem.

But here is an interesting paradox: a still more abstract physical theory proved to be able to make more concrete predictions concerning possible Tunguska traces. I mean the hypothesis explaining the Tunguska phenomenon as a collision of Earth with an asteroid consisting of the so-called "mirror matter." This idea was put forward in 2001 by the Australian physicist Robert Foot. So, what does this strange combination of words – mirror matter – mean?

It was in 1956 that American physicists Tsung-Dao Lee and Chen-Ning Young discovered that electrons and neutrinos arising when a neutron decays are always "left-handed." An observer toward whom these elementary particles flew would see them rotating clockwise. The scientists were awarded a Nobel Prize for their discovery, but the physical research community got upset – why such asymmetry? No physical law prescribes this specific order of things. There is good reason to believe that "right-handed" particles can also exist, and these were later called "mirror particles." But where should they be searched for?

It had already been established that, apart from ordinary elementary particles – the electron, proton, neutron, and others – there also exist antiparticles: positron, antiproton, antineutron, antineutrino, and so on. These had been predicted in the 1920s by the famous British physicist Paul Dirac from a different line of reasoning, and the first antiparticle (positron) was discovered experimentally in the 1930s. So, the Soviet physicist Lev Landau (that very man who explained to Alexander Kazantsev the physical principles of atomic explosion) had supposed that the hypothetical mirror particles and the well-known antiparticles are the same thing. Physicists agreed, and the physical world became symmetrical again. However, this situation did not last long. In 1964 the young American physicists James Cronin and Val Fitch, two future Nobel laureates, proved experimentally that Landau's hypothesis was wrong, and the asymmetry in the decay of subatomic particles still existed.

So physicists had to look for different candidates for the title of "mirror particles." And such candidates were found – or rather theoretically predicted – two years later by Soviet physicists Isaak Pomeranchuk, Lev Okun, and Igor Kobzarev. Their hypothetical mirror particles differed from antiparticles in that they could interact with ordinary subatomic particles only by gravitation. If a neutron and an antineutron collide they are annihilated, whereas colliding neutron and "mirror neutron" particles will simply "ignore" each other. But between themselves mirror particles interact absolutely normally and therefore there can exist cosmic bodies and systems consisting of mirror matter – mirror galaxies, stars, and planets. What is more, even in our galaxy there may exist double stars, one component of which consists of normal matter and the other of mirror matter – at least theoretical physics makes this possible.

As sometimes happens in science, the idea proposed by Pomeranchuk and his colleagues was discussed in the science community and then forgotten for 20 years. Its renaissance occurred in the 1980s and especially in the 1990s, when astronomers and cosmologists concluded that so-called "dark matter" (or "hidden mass") must be present to explain the gravitational dynamics of the universe. Astrophysicists have found that more than 95% of matter existing in our universe should constitute the invisible hidden mass, which is detected only by its gravitational influence on stars and galaxies. The origin of this mass remains unknown, but the hypothetical "mirror matter" is a very good candidate for this position. It fits well the two main characteristics of dark matter. First, it cannot be seen – because mirror photons emitted by mirror matter do not interact with normal matter. At the same time, mirror matter does interact with normal matter gravitationally, that is, through the omnipresent force of attraction between any particles of matter in the universe.

Dr. Robert Foot, who supported the idea of dark and mirror matter, disagreed, however, that the normal world and the mirror world would be almost completely separated from each other. He supposed that apart from the gravitational interaction between them, there could exist one more type of interaction – directly between photons and mirror photons.[12] If there is no such interaction, then even large asteroids consisting of mirror matter may pass through the atmosphere of our planet and even through the planet itself, not disturbing anything and remaining therefore unnoticed. But the situation changes considerably if photons and mirror photons do interact.

In this case, upon entry of a mirror space body into the atmosphere a drag force arises that swiftly heats the body. A large chunk of mirror ice on course to hit Earth with a cosmic velocity would melt at an altitude some 5–10 km, which corresponds well with the altitude of the Tunguska explosion. While it is melting and being dispersed in the air, the atmospheric drag force would sharply increase and the body would explode, releasing its kinetic energy into the atmosphere.[13]

So, if the TSB was indeed a mirror asteroid, the absence of the ordinary meteoritic substance in the Great Hollow becomes understandable. In addition, some fragments of mirror substance can still be expected, if it was not *too* volatile. Who knows, perhaps these fragments are still awaiting someone to discover them on the site. According to such physicists as Robert Foot and Zurab Silagadze, they could be found there. True, the task of digging them out may become much too difficult. As it follows from the theory, even if the mirror matter can interact with normal matter this interaction is very, very weak.

Needless to say that the "mirror hypothesis" of the TSB nature is not so much an astronomical conception as a purely physical one, emerging from a "frontier area" of physical science. Astronomers, especially meteorite specialists, have been accustomed to less-sophisticated theories and therefore they feel instinctive doubts about such considerations. For example, when Robert Foot attempted to explain some peculiarities of craters that had been photographed by the space probe *NEAR Shoemaker* on the surface of asteroid Eros in 2000 as resulting from collisions with mirror asteroids, astronomers just shrugged their shoulders. Thus, it should hardly be expected that the mirror model of the Tunguska phenomenon will soon take the leading place in this field of investigations – even though it's rather promising. But at present this model looks too far-reaching – "too cosmic."

An opposite approach to the Tunguska phenomenon – that is, attempts to declare it a purely terrestrial event – is evoking in the general public (if not among the specialists) much greater interest, or at least is understood better than the above microphysical theories. Really, why should all these researchers cudgel their brains over all that kind of rot, trying to associate enigmatic traces in the taiga with unknown parameters of a fantastic cosmic body? What if there was no cosmic body at all? What if it was just an unusual earthquake, or something like that?

The most active – and the most well-known – partisan of the purely "terrestrial" approach to the Tunguska problem is Dr. Andrey Olkhovatov, who in 1997 published in Russia a book with a provocative title: *The Myth About the Tunguska Meteorite. The Tunguska Event of 1908 as a Mundane Phenomenon*. As its author openheartedly informs the readers, "The idea about the Tunguska phenomenon as a product of tectonic processes came to my mind in the late 1980s, when I happened to read a couple of popular science books about earthquakes. Although I had never studied the Tunguska problem before, I was astonished by the similarity between Tunguska eyewitness accounts and those of witnesses of some earthquakes."[14]

Many papers by Olkhovatov have been published in various Russian and foreign periodicals. So, what data is Olkhovatov considering? Generally, these are the same well-known facts discovered by Tunguska investigators: no material substance of the TSB has been found; optical atmospheric anomalies had started several days *before* the TSB fall; neither the meteorite fall nor that of a comet core can explain the thermoluminescence and the paleomagnetic anomaly, the post-catastrophic accelerated growth of trees, genetic mutations at Tunguska, and so on.[15]

And what was Olkhovatov's conclusion from these facts? Very simple, even if not very logical: there was no TSB at all. So what was there instead? Judging from the literature, both advocates and opponents of Olkhovatov's viewpoint believe that it was an ordinary earthquake that caused the Tunguska phenomenon. This, though, was not his hypothesis, which is more exotic. According to Olkhovatov, there occurred at Tunguska the so-called "natural non-local explosion" (NNLE) – a new, previously unknown type of seismic activity "which is something other than an earthquake, even if rather similar to it."[16] That is, we are dealing here with an underground variety of exotic cosmic body that has never been observed by geophysicists and seismologists – neither before nor after the Tunguska explosion. Olkhovatov did quote in his works the descriptions of luminous formations sometimes appearing in the atmosphere before earthquakes or accompanying these. But in fact, these phenomena are essentially different. Neither their scales nor consequences are even comparable.

By the way, Olkhovatov refrains from describing the mechanism of NNLE in any detail, leaving it an enigma on its own. And it is so easy to explain one enigma via another one. But has an NNLE ever been recorded releasing the energy of 50 Mt of TNT? Olkhovatov's reference to the so-called Sasov explosion that took place in the Ryazan region of Russia on the night of April 12, 1991, has nothing to do with the case. Its magnitude was about 300 tons of TNT, that is, 100,000 times weaker than the Tunguska explosion; nonetheless, Olkhovatov calls it "mini-Tunguska." Why not call it "micro"? Yes, earthquakes are from time to time accompanied by strange light phenomena, but this does not mean that all strange light phenomena are generated by earthquakes or a fantastic NNLE. Incidentally, according to Olkhovatov, poltergeist is also an NNLE[17] as well as ball lightning. Then perhaps we should attribute the whole Tunguska phenomenon to a gigantic poltergeist? That would have been a truly original hypothesis!

Unfortunately the "purely mundane" origins of the Tunguska event are enthusiastically received and supported by those readers who have a poor grasp of the data collected during the century of Tunguska investigations. (When, some years ago, Andrey Olkhovatov described his hypothesis on his website, the web server was overloaded by people wishing to "know the final solution of the Tunguska enigma.") As for the specialists in the Tunguska problem, they find themselves in an unenviable position. Discussions with absurd statements could last infinitely – and lead nowhere. For example, Vitaly Bronshten, in his very substantial book *The Tunguska Meteorite*, somewhat perplexedly informed his readers: "But there had been a TSB, indeed!" This hardly convinced Olkhovatov's supporters.

To be truthful, Andrey Olkhovatov's contribution to the problem of the Tunguska meteorite closely resembles a sudden intervention of a passerby into a discussion group of geophysicists about the shape of our planet. The specialists are debating which dimensions of the globe should be considered as sufficiently precise and which is the polar radius of Earth, whereas a new participant appears and states aloud: "What are you quarreling about? Earth is flat and standing on three whales! This is the model that gives the best fit to all data collected by now!"

At the same time, Olkhovatov does notice very well those nuances and peculiarities of the Tunguska phenomenon that cannot be explained by its cometary and meteoritic models. For instance, he proves convincingly that the "fiery ball" flying over the Great Hollow could explode only due to its internal energy and not due to its energy of motion (confirming thereby Alexey Zolotov's and Maxim Tsynbal's conclusions – if they required any additional confirmation) and demonstrates the complicated character of the TSB's flight path... Well, and...? It is self-evident that the Tunguska phenomenon is full of various enigmas, but to explain them away with the help of a mythical NNLE eruption does not mean to work out the Tunguska problem. Rather, it means to muddle matters.

Now, why are the majority of "exotic" Tunguska hypotheses, both mundane and cosmic, inadequate? Why, after all, cannot an extraordinary event be explained by an extraordinary hypothesis? Well, perhaps it can be and even should be. But these hypotheses are either ignoring well-established facts or cannot generate any predictions through which it would become possible to verify them. Sometimes it is even both of these. Of course, ignoring facts is blameworthy, but it is only rarely that necessary attention is paid to the inability of a hypothesis to be testable via verifiable predictions. However, this self-test is the most important component of the whole scientific method of cognition. It is far from sufficient to say that, for example, the cometary hypothesis cannot explain some traits of the Tunguska event, whereas some "super-NNLE" can. The scientist still has to prove that it is *only* the "super-NNLE" that can account for this event.

That is why attempts by "conservative" advocates of the cometary-meteoritic TSB models to build advanced schemes of the Tunguska event, involving a comet core or a stony asteroid, should not be rejected out of hand. Recently, a group of Tunguska researchers from St. Petersburg – Dr. Henrik Nikolsky and Edward Schultz at the Institute of Physics of St. Petersburg University, and Professor Yury Medvedev at the Institute of Applied Astronomy – attempted to calculate a trajectory for the Tunguska comet that would best fit all the empirical facts. They suggested that the TSB was a fragment of Comet Encke, which had been discovered in 1786 and is revolving around the Sun with a period of just 3 years and 4 months (the shortest known cometary period of revolution). In December 1907, when approaching its perihelion (that is, the point in its orbit where it is nearest to the Sun), this comet broke up into several large pieces. One of these pieces, the St. Petersburg scientists believe, approached Earth, touched its upper atmosphere, and decelerated, after which it was caught by the gravitational field of our planet and became its temporary satellite. This was a cosmic body some 400 m across and with a mass of about 30 million tons. It made its first revolution around Earth in an orbit with an apogee – the maximal distance from the planet – of 60,000 km (six times closer than the Moon) and a perigee – its minimal distance – just 40 km distant. This was over Antarctica. Naturally, at such a low perigee the cosmic body would have been slowed down by air resistance, and so its altitude decreased with every orbit. Also, flying through increasingly denser layers of the atmosphere, the body's substance began to burn up. The Tunguska comet made three complete revolutions around Earth, losing half of its mass and producing atmospheric anomalies that, as we know, had started as early as June 27, 1908. When over Europe, the comet disturbed the geomagnetic field, the perturbations of which were recorded by Professor Weber in Kiel, Germany. By its fourth revolution around Earth, the TSB's speed was already less than that needed to keep it in orbit, and its altitude on its fourth incomplete circuit was just 100 km. Moving toward Tunguska along the 101st eastern meridian, somewhere before Kezhma, the TSB broke apart into several fragments.[18]

Each of these fragments was burning up and intensively evaporating, the whole volume of the explosive cloud reaching 200 km^3. And when the speed of the TSB fragments diminished to a couple of kilometers per second, the cloud detonated, its explosion lasting about five seconds. The blast wave hit the taiga, leveling trees. Two seconds later, scorching gases fell upon Earth's surface, burning the trees, bushes, and moss in the Great Hollow. Lesser vapor clouds, formed by other TSB fragments that followed the first one with intervals of several seconds, exploded as well, additionally devastating the taiga. Chemical products of these explosions were ejected into the upper atmosphere, where they brought about optical anomalies and the local geomagnetic storm. As for the lumps of cometary ice that had survived the explosions, these fell to the Southern swamp and melted there.[19]

One cannot but notice a close similarity between this hypothesis and the hypothesis proposed by Maxim Tsynbal and Vladimir Schnitke that was described above. The idea of the vapor cloud explosions has been taken from there (with due references, of course). Naturally, all its shortcomings have remained intact; but the hypothesis of the "orbital comet" goes far beyond the limits of the former hypothesis. Its essential advantage is the authors' desire to take into account as many Tunguska traces as possible. They have even paid attention to an unusual atmospheric glow that had been observed in Antarctica, near the Erebus volcano, just a few hours before the Tunguska event, by Professor T. W. Edgeworth David, the scientific chief of the Anglo-Australian Antarctic expedition of 1908.[20] According to the Russian researchers, this glow was associated with the Tunguska comet flying past Mt. Erebus, in the lowest part of its orbit. But the key advantage of their scheme is that they propose fairly rational explanations for a whole group of phenomena accompanying the Tunguska event – not just for a couple of them.

First, this scheme proposes an explanation for probably the most enigmatic precursor of the Tunguska phenomenon, the Weber effect – strange perturbations of the geomagnetic field recorded by Professor Weber in Kiel, Germany. As the St. Petersburg scientists state, it was generated by the orbital motion of the TSB. Second, we can now trust the reports of those eyewitnesses from Kezhma, some of which saw a flying body to the east from the village and the others to the west. These were the separate fragments of Comet Encke. The accounts of the Evenks about several powerful explosions and a strong quake *before* the first explosion – which, as we know, awoke Chuchancha and Chekaren, who were peacefully sleeping in their chum – also become better understandable. Nikolsky and his coauthors believe that this quake was produced by the fall of a "huge icy fragment of the TSB" into the Southern swamp. It is suggested that the local geomagnetic storm could have been due to an ejection of chemical products from the explosions of vapor cloud in the ionosphere. These products, weighing tens of millions of tons, made a "hole" in the ionosphere and disturbed ionospheric electric currents, which affected the geomagnetic field. But what is especially important, the anomalous atmospheric phenomena that took place both before and after the Tunguska explosion do also obtain a natural explanation. Before the explosion they were due to the loss of cometary substance during the TSB's orbits and afterward due to the ejection of the explosion products into the upper atmosphere.

As for genetic mutations and anomalies of thermoluminescence, the researchers accepted the scheme developed by Tsynbal and Schnitke – a breakout of the high-energetic ultraviolet radiation through the ionospheric hole. Equally, they have agreed with Sokrat Golenetsky and Vitaly Stepanok that it was the "cometary fertilizer" (that part of the TSB substance that got into the soil – not in the upper atmosphere) that promoted the accelerated restoration of the forest at Tunguska.

A beautiful hypothesis indeed! A clever, well-developed, and flexible one. Calculations of possible capture of the TSB by the terrestrial gravitational field and its subsequent orbital maneuvers have been made at a high professional level. But once again – where is the TSB substance? "It dissolved in the Southern swamp." Such an explanation looks very strained. To dissolve leaving no trace the cometary ice must have been extremely pure. This contradicts the recent astronomical data. Besides, the Weber effect – the strange regular oscillations of the geomagnetic field – occurred on June 27, 28, and 29, 1908, exactly 24 hours apart. How could the "orbital comet," whose period of revolution never exceeded 10 hours, generate the Weber effect? And also, how could a fragment of the icy comet core, flying at an altitude of tens of thousands of kilometers, perturb the geomagnetic field so much as to be recorded in Kiel?

And last but not least, it is evident that products of a chemical explosion, even though very powerful but devoid of any radioactivity, could not give rise to a local geomagnetic storm lasting five hours. At best, a geomagnetic disturbance, brought about by the vapor cloud explosion of the TSB, would have lasted several minutes, until all electric charges in the fiery ball had been neutralized.

Nonetheless, despite all these defects, at present it is the hypothesis by Dr. Henrik Nikolsky and his colleagues that may be considered as the most advanced version of the cometary explanation of the Tunguska phenomenon. Perhaps, its further development will open way to new progress in Tunguska investigations.

Thus, in previous chapters we have described 10 traces that remained after the Tunguska event – from the radially leveled forest and light burn of the Tunguska vegetation to genetic mutations and indications of radioactivity. We also considered eyewitness reports – which should certainly be taken into account when searching for the correct solution of this problem. In this chapter we have considered 10 hypotheses, whose authors are trying to explain these traces and to find out the nature of the Tunguska phenomenon – from a comet and a stony meteorite to the "natural non-local explosion" and an asteroid consisting of "mirror matter." Each of these hypotheses meets with considerable difficulties when trying to account for all peculiarities of this phenomenon, and therefore science does not possess as yet the correct theory.

Does this mean that the efforts of scientists who, during many decades, were putting forward and developing Tunguska hypotheses were in vain? Far from it. In a preceding chapter we saw that from the 1960s the scientific community, having made sure that it was impossible to take the "Tunguska fortress" by storm, went over to a more systematic siege. Specialists in various scientific disciplines have built around this fortress, so to say, a system of trenches helping them to work out their theories and to check if they correspond to known Tunguska facts. And this siege has borne some fruit. A map of the fortress, with its 10 "bastions" – traces of the Tunguska phenomenon – is now available. A circle of the "science army" around these bastions gradually becomes tighter, preparing for the final assault. Hypotheses and theoretical models of the Tunguska phenomenon may be compared with siege guns: success of the future assault depends, first of all, on their quality and caliber. The experience of this long siege has shown that a great many of these siege guns are, alas, ineffective against the walls of the Tunguska fortress, though some of them may still be useful.

So, which of the "siege-guns" have been sent to a melting furnace or at least withdrawn from service? First of all are the fringe hypotheses that suggest there was no cosmic body over Tunguska and that the phenomenon is explainable in terms of ball lightning, an explosion of marsh gas, an unusual hurricane, or an unusual earthquake. Eyewitness reports may not be that exact, but the very fact that they exist does convincingly testify that there was a cosmic body flying in the atmosphere for about 1,000 km. Also, the thoroughly investigated area of leveled forest and that of the light burn prove that the body exploded in the air over the Great Hollow at an altitude of 6–8 km. The iron meteorite hypothesis has also been refuted: no one reported a dense tail of iron particles behind the flying TSB and no pieces of meteorite have been found at Tunguska.

As for hypotheses explaining the Tunguska catastrophe by the arrival from space of such exotic objects as a black hole, solar plasmoid, or cosmic snowflake, these have remained just "initial conjectures" and have not become scientific hypotheses in the strict sense of this term. These conjectures either contradict well-established empirical facts or cannot generate any verifiable predictions. Regarding Alexander Kazantsev's spaceship hypothesis, which played a very important part in the history of the Tunguska problem, its progress has practically stopped. Having predicted some important facts: the overground character of the Tunguska explosion, the lack of meteoritic substance in the Great Hollow, traces of radioactivity and genetic mutations, this hypothesis ceased to evolve and lost – perhaps temporarily – its "predicting potential." But what, after all, may it predict if we have no idea of the searched-for object – an extraterrestrial spaceship?

On the other hand, the starship hypothesis does explain more easily and convincingly than a comet or a stony meteorite such aspects of the Tunguska event as the local geomagnetic storm, the rare earth anomaly in Tunguska soil (which can have no relation to small cosmic bodies), the anomaly of thermoluminescence, and especially possible maneuvers of the TSB in its flight to the Great Hollow. To return this hypothesis to the leading place in Tunguska studies that it had several decades ago, its supporters would have to look for material remnants of the TSB. But for the time being, it is a comet and a stony asteroid that are generally considered the chief candidates in the Tunguska mystery, even though each of these has its own serious drawbacks. But many scientists are certain that to solve this problem means to choose between these two hypotheses.

This choice has turned out to be more difficult than could have been imagined several decades ago. Somehow, the properties of either of these small cosmic bodies fall short of explaining all well-established facts relating to the Tunguska phenomenon. A comet core could not have penetrated the atmosphere so deeply, whereas a stony asteroid would have left a large amount of rocky substances in the Great Hollow. One can imagine a "cometary fertilizer" accelerating the growth of the taiga vegetation, but hardly a fertilizer consisting of meteoritic rock. On the other hand, a "radioactive meteorite" (not composed of pure uranium-235, as Alexander Kazantsev had assumed, but at least containing some radioactive elements that might have been responsible for the radiation effects discovered at Tunguska) looks somewhat more acceptable than a "radioactive comet core."

But as a whole, it seems that the real TSB must have possessed altogether the properties of a stony asteroid and those of a comet core. It had to be at least as strong mechanically as a stony asteroid to attain the altitude of 6–8 km before it disintegrated. It also had to contain still less hard substances than a normal comet core has. And finally, there must have been in the TSB something that made possible its detonation over the Southern swamp.

Very contradictory requirements, one has to admit! Perhaps, a mirror asteroid could have contained all the necessary traits, but as said above, for meteor specialists this hypothesis seems too alien. Nevertheless, however strange it may sound, the exotic mirror model is rational, theoretically substantiated (physicists are persistently looking for mirror matter), and verifiable, at least in principle.

Albert Einstein has wonderfully described the main properties of a truly good scientific theory: it must possess, on the one hand, "external confirmation" and, on the other hand, "inner perfection." In other words, a theory is good when it accounts for all well-established facts associated with it and when it does that from a minimal number of initial suppositions.[21] Of all 10 Tunguska hypotheses that we have considered, it is probably the "orbital comet" by Henrik Nikolsky and his colleagues that possesses the best external confirmation – even despite all its weak points. At least, its authors are trying to cover all facts accumulated in the Tunguska file. But it lacks inner perfection. The complexity of the scheme, developed by the St. Petersburg scientists, rather hints at gravitational maneuvers of an extraterrestrial spaceship than at a simple comet. As for the best inner perfection, this is found in the mirror asteroid model, though its external confirmation leaves much to be desired. Such a sharp contradiction between the two perhaps brightest contemporary hypotheses about the nature of the TSB seems to suggest that their chances of becoming the last word in the long controversy are not good.

Certainly, one can understand the broad audience that is inclined to believe every new idea about the nature of the Tunguska phenomenon – independent of the level of its justification. The infinite vacillations of meteor specialists between a stony asteroid and an icy comet core can hardly evoke enthusiasm. A 100 years of the history of the Tunguska problem – and 80 years of active investigations – is a sufficiently long period for the nonspecialists to become irritated with the progress made.

There is sad truth in this irritation. But who could have expected 80 years ago that the Tunguska problem would turn out so difficult, and especially so multidisciplinary? To find a correct explanation for every Tunguska trace is a challenging task, but still more challenging is combining these explanations into a unified picture. A biologist studying genetic mutations in Tunguska pines and a physicist investigating the local geomagnetic storm that started soon after the explosion are speaking very different scientific languages, and it is difficult for them to understand each other. As a rule, the biologist has a very general idea of the ionosphere, as the physicist has of the molecule of DNA, so how can they find a common ground for investigating the Tunguska phenomenon – or even for discussing it?

And such difficulties are constantly emerging before Tunguska researchers. So, perhaps the scientists besieging the Tunguska fortress have huddled into their "disciplinary trenches" somewhat too early? Yes, it is safe in these trenches, and one can build there highly professional schemes of the enigmatic event that occurred at Tunguska a century ago; but communications between different trenches are bad and attempts to summon up the existing scientific forces regularly fail. Luckily, there is a way out of this situation. We should retreat a little, have a better look at the besieged fortress, and try to build its model demonstrating the Tunguska phenomenon as it was. Then the real picture of the phenomenon would emerge not obscured by theoretical veils. This is what we will try to do in the next chapter.

Notes and References

1. Petrov, G. I., and Stulov, V. P. Motion of large bodies in planetary atmospheres. – *Kosmicheskiye Issledovaniya*, 1975, Vol. 13, No. 4, p. 594 (in Russian).
2. Bronshten, V. A. *The Tunguska Meteorite: History of Investigations*. Moscow: A. D. Selyanov, 2000, p. 148 (in Russian).
3. See, for example, D'Allesio, S. J. D., and Harms, A. A. The nuclear and aerial dynamics of the Tunguska event. – *Planetary and Space Science*, 1989, Vol. 37, No. 3.
4. Its crushing strength is 2.5 MPa. See Tsynbal, M. N., and Shnitke, V. E. A gas–air model of the Tunguska comet explosion. – *Cosmic Matter and the Earth*. Novosibirsk: Nauka, 1986, p. 186 (in Russian).
5. See Sekanina, Z. The Tunguska event: no cometary signature in evidence. – *Astronomical Journal*, 1983, Vol. 88, No. 1.
6. Grigoryan, S. S. About the nature of the Tunguska meteorite. – *Reports of the USSR Academy of Sciences*, 1976, Vol. 231, No. 1; Grigoryan, S. S. On the motion and destruction of meteorites in planetary atmospheres. – *Kosmicheskiye Issledovaniya*, 1979, Vol. 17, No. 6; Bronshten, V. A. On the dynamics of destruction of large meteoroids. – *Kosmicheskiye Issledovaniya*, 1985, Vol. 23, No. 5 (in Russian).
7. See Bronshten, V. A. A plenum of the commission on comets and meteors. – *Priroda*, 1951, No. 11 (in Russian).
8. Solyanik, V. F. The Tunguska catastrophe of 1908 in the light of the electric theory of meteor phenomena. – *Interaction of Meteoritic Matter with the Earth*. Novosibirsk: Nauka, 1980 (in Russian).
9. Nevsky, A. P. Explosions of meteorites due to electric discharge as a global danger for civilization. – *Tungussky Vestnik*, 1997, No. 6 (in Russian).
10. Bronshten, V. A. *The Tunguska Meteorite: History of Investigations*, p. 242 (in Russian).
11. Jackson IV, A. A., Ryan, Jr., M. P. Was the Tungus event due to a black hole? – *Nature*, 1973, Vol. 245, No. 5420.
12. The so-called photon–mirror photon mixing, see Foot, R. *Acta Phys. Polon.*, 2001, B32; Foot, R., and Yoon, T. L. *Acta Phys. Polon.*, 2002, B33.
13. For details see Silagadze, Z. K. Tunguska genetic anomaly and electrophonic meteors. – *RIAP Bulletin*, 2006, Vol. 10, Nos. 1 & 2.
14. Olkhovatov, A. Y. *The Myth about the Tunguska meteorite. The Tunguska Event of 1908 as a Mundane Phenomenon*. Moscow: Association Ecology of the Unknown, 1997, p. 3 (in Russian).

15. Ibid., pp. 6–7.
16. Ibid., p. 6.
17. Ibid., p. 32.
18. The authors do however admit that the Tunguska space body could have been a group of separate bodies from the very beginning, before entering the terrestrial atmosphere.
19. See Nikolsky, H. A., Medvedev, Y. D., Schultz, E. O. A balanced model of the Tunguska phenomenon. – *Proceedings of the Conference "Centenary of the Tunguska Cometary Body."* St.-Petersburg, 2008 (in Russian).
20. See Steel, D., and Ferguson, R. Auroral observations in the Antarctic at the time of the Tunguska event, 1908 June 30. – *Australian Journal of Astronomy,* 1993, Vol. 5, March.
21. See Kuznetsov, B. G. *The Eisteinian Essays.* Moscow: Nauka, 1970, pp. 192–193 (in Russian)

11. The Theory is Dead: Long Live the New Model

Dozens of books and hundreds of articles have been published about Tunguska. This subject has appeared in academic journals as well as in popular scientific and fringe periodicals. Several dissertations for degrees have been defended and many papers have been read at conferences. Researchers have collected a wealth of evidence of the Tunguska catastrophe, and this information has been thoroughly analyzed. But strange though it may seem, nowhere can you find a complete and objective reconstruction of the Tunguska event. As a rule, having depicted almost exactly some aspects of Tunguska, the author of an article or a book immediately jumps to the description of the event – *how it should have looked from the viewpoint of the hypothesis that this author is supporting.* For example, "The core of a small comet came flying into the terrestrial atmosphere with the speed of about 30 km/s and began to intensively evaporate," or "a stony asteroid with a mass of 300,000 tons, gradually collapsing under the action of the powerful air resistance, was moving at an enormous velocity over Siberian wastes." Always a purely "theoretical" picture. "Here is how the phenomenon must have looked, and those Tunguska traces, which do not correspond to the proposed picture, have nothing to do with this event." So say these authors.

Undeniably, to discriminate between information bearing on the problem at hand and unrelated information is an important stage of scientific investigation. The trouble is, however, that some "theoretically irrelevant facts" may turn out to be very relevant, especially when we are investigating a natural phenomenon and not just analyzing results of an experiment that was carried out in a laboratory. Experiments are the basis of the scientific method of cognition because they are conducted in artificially clean conditions. Due to this, their results may be considered as reliable and precise. But when we are working with an out-of-laboratory phenomenon, whose origin and nature are a priori unknown, we are at risk,

V. Rubtsov, *The Tunguska Mystery*, Astronomers' Universe,

DOI 10.1007/978-0-387-76574-7_11,

when filtering out "useless data," to throw away the essential together with the inessential.

So, let's forget for a moment about theories and pay attention to empirical facts. After all, it was the "unpleasant facts" (such as the overground explosion of the Tunguska space body) that have provided the basis for investigations at Tunguska covering many years. Even though the meteor specialists have after all succeeded – not without difficulty – in finding a theoretical explanation of the overground meteorite explosion with the help of the theory of the swift fragmentation that we described in the previous chapter, the credit for this success goes more to Alexander Kazantsev than to these specialists. If he had not paid attention to this subject, why should anybody have attempted to explain it? Most probably, every astronomer would have believed even now that a meteorite may explode only when striking a hard surface.

Of course, a "purely empirical" image of the Tunguska phenomenon cannot be absolutely unambiguous – otherwise the Tunguska mystery would have been solved long ago. It would have been enough to take the existing elements of this jigsaw and assemble from them an evidently correct picture. But *sufficiently* definite and sufficiently *accurate* it must be. We have at present a lot of important empirical data, collected in the swamps and copses of the Great Hollow, which can be used for this purpose.

The traces of the Tunguska event that were considered in previous chapters are its direct and indirect consequences, providing valuable information about various parameters of the Tunguska explosion, the dynamics of the TSB flight, and the TSB itself. To be revealed, this information requires effort and persistence on our part. Let's therefore try to reconstruct these parameters and traits. But first we should agree upon an important precondition, that is, not to start work by separating the sheep from the goats and bringing in a verdict, which has often happened in the past. Let's put our trust in the results of long studies conducted in the Great Hollow and eyewitness testimonies collected in the villages surrounding it. Also, keep in mind that we are not trying to answer here the question about the nature of the Tunguska phenomenon. We are just describing it as objectively as possible.

Will the final reconstruction be comprehensive? Not necessarily. We cannot be sure that science at present possesses *all* the facts needed for a complete reconstruction of the Tunguska event. But our reconstruction will definitely be much more complete – and more reliable – than theoretical descriptions of this event, based on hypotheses rather than on facts. Of course, it might have happened that by a miracle, that is "intuitively," the researcher could hit upon the correct answer to the problem. In this case, the theory would certainly have made it possible both to correctly reconstruct the Tunguska event and to convincingly explain the traces it has left. But this hasn't yet been possible. This is why we have to use another method to solve the problem – a purely empirical one. We will remove hypothetical schemes of the Tunguska event and simple-mindedly follow the facts we have. No guesses – just objectivity, empiricism, and taking into account all reliable data.

To begin with, let's remember which empirical data we possess at present. There are three large Tunguska traces: the area of the leveled forest, the light burn, and the local geomagnetic storm. And there are seven lesser traces: genetic mutations of plants, insects, and humans; an accelerated growth of the Tunguska vegetation; fluctuations of the radioactive background in the Great Hollow and a radioactive contamination in the tree rings dated 1908; the thermoluminescence anomaly; the paleomagnetic anomaly; the Weber effect; and, importantly, the geochemical anomalies in Tunguska soil and peat. In addition to these traces we have got instrument recordings of seismographs and barographs, as well as a great number of testimonies of eyewitnesses who saw the flight and explosion of the TSB. Also, there are detailed descriptions of the atmospheric optical anomalies – both preceding the event and following it (the latter being especially intensive). So, there is extensive data available. Let's agree, this is far from naught!

The area of the leveled taiga may be considered as the very foundation of the Tunguska problem. Had there been no forest leveling (which could have been the case if, say, the TSB had exploded at an altitude of 50 or more km), then nobody would have ever bothered to study anything in the epicenter of the explosion. Some 30 million leveled trees do, therefore, have some significance. This is the "main" Tunguska trace, not the "first among equals" but the very first. That it was mapped before the trees had rotted is probably the main achievement of the Independent Tunguska Exploration Group and of Wilhelm Fast personally.

The light burn is also very informative evidence of what took place in 1908 over the Great Hollow. The share of the light emission in the whole radiation from the fiery ball of the Tunguska explosion was for a long time considered as the critical parameter for the choice of the hypothesis explaining its nature. If it was high enough, the explosion must have been nuclear; if not, then nonnuclear. It has since been proved that such an option was invalid, because if a meteorite or a comet core flying through the atmosphere was heated to a high-enough temperature, the share of the light emission in its radiation would be comparable to what would be produced by a nuclear explosion. Moreover, a vapor cloud explosion (definitely a chemical one), having a relatively low temperature – just 2,000–3,000°C – generates a powerful stream of infrared radiation that could also have left the observed imprint on Tunguska vegetation.

The local geomagnetic storm several minutes after the Tunguska explosion is perhaps its most unusual consequence. The only model that convincingly explains it is the model in which this effect was produced by the ionizing radiation of the fiery ball of the Tunguska explosion. Attempts to explain this geomagnetic storm via the action of the blast wave or the ballistic shock wave from the flying TSB on the ionosphere have failed. But by admitting that there was ionizing radiation, it would be necessary to consider a difficult question: where did this ionizing radiation come from? Few Tunguska investigators are daring enough to go so far as to accept that the Tunguska explosion was accompanied by nuclear reactions – even though such reactions would not necessarily imply an extraterrestrial visit.

Fortunately, even though we do not yet know what the TSB was, we know fairly well how the Tunguska event occurred. Therefore, an examination of the known facts can lead us to a justified conclusion about this phenomenon. To start with, judging from eyewitness testimonies, at least one space body of enigmatic origin traveled through the atmosphere some 1,000 km before it exploded over the Great Hollow. ("At least" means that there might be more than one flying object, but there was a space body, in any case. Fantasies about unusual hurricanes and earthquakes have remained in their proper place – in the 1920s.) This is both the most general picture of the Tunguska event and the starting point from which we can proceed further.

Is this too little? Not at all. For instance, the great length of the atmospheric path of the TSB tells us that the space body was flying at a small angle to Earth's surface. This angle could not have exceeded 10–15°; otherwise, the altitude at which the TSB began to emit light would have been too great.[1] And as we already know, the body could not have been in a sharply increasing descent in the final stage of its flight, or else it would have been destroyed by the g loading.

Now what was the TSB's velocity? After processing the eyewitness testimonies, the ITEG scientists have established that the space body was flying over Siberia for about 5 min.[2] Taking into consideration the distance it had covered – some 1,000 km – we can assess its *average* speed to have been about 3 km/s. Of course, this is just a tentative estimation, but it's not devoid of interest because meteorites usually fly into the atmosphere at much greater velocities.

But as for the speed of the TSB at the end of its path over the Great Hollow, it can be determined more precisely and could not exceed the speed of a hypersonic aircraft. Otherwise the body, flying in a flat trajectory, would have left in the leveled forest a more pronounced trace of its ballistic shock wave than it did. A steep TSB trajectory and great velocity (tens of kilometers per second), which appear in many Tunguska hypotheses, are only there because these figures are necessary to justify an amount of kinetic energy that would be needed for a thermal explosion or a swift fragmentation of the body. But a flat trajectory and a low final speed (a couple of kilometers per second at best) are what the empirical facts indicate. If the TSB was seen at a distance of 800 km from the epicenter (in fact, it was seen at distances of more than 1,000 km!), and its flight lasted some 5 min, it means that its trajectory *had* to be flat and its speed low. By the way, the low speed of the TSB eliminates the problem of its strength over which the cometary hypothesis has stumbled. With a low speed, even such a fragile object as a comet core could have reached the Southern swamp intact.

Thus, the TSB had a flat trajectory and a low velocity. Not a steep trajectory and a high velocity. Therefore, calculations and models of the Tunguska event based on a steep TSB path and great velocity may be of interest as mathematical constructions, but they have nothing to do with the Tunguska event. Also, the share of the energy from the ballistic shock wave in the whole of the energy released at Tunguska was negligible. All fifty Tunguska megatons are "megatons of explosion" and not "megatons of motion."

The strict radiality of the area of the flattened forest testifies that there was only one powerful Tunguska explosion. If there had been another explosion whose magnitude was comparable to the first one (even after the trees had fallen), this radiality would have been broken and its blast wave would, most probably, have been recorded by seismometers and barographs in Russia and elsewhere. Nothing of this sort occurred, and therefore we can say with certitude that there was no other equally powerful Tunguska explosion – just one. And from the seismograms the time of this explosion has been determined to within ten seconds.[3] Also, as explained earlier, it has been established that the Tunguska explosion occurred in the air at a relatively high altitude – between 6 and 8 km, judging from the diameter of the zone of standing trees at the epicenter of the explosion. Some additional estimations of this altitude were made with the help of seismograms and barograms, and they do not contradict this assessment. And judging from the area of destruction and the energy of aerial and seismic waves, we can accept that the magnitude of the main explosion was several dozens of megatons.[4]

It is, however, probable that apart from the main explosion there were at least two low-altitude and less-powerful explosions. It was Leonid Kulik who had discovered their epicenters on aerial photographs, and later his conclusions were confirmed by Siberian scientists. Dmitry Demin and Sergey Simonov found additional proof when analyzing the subtle structure of the area of the leveled forest, and Sokrat Golenetsky with Vitaly Stepanok discovered one of these local epicenters when examining an elemental anomaly.[5] Remember also the testimony of the Evenk brothers Chuchancha and Chekaren, who confirmed that there were *several* explosions: "We saw another flash of light while thunder crashed overhead followed by a gust of wind that knocked us down. Then Chekaren cried out: 'Look up!' and stretched his hand upward. I looked and saw new lightning and heard more thunder."[6] These less-powerful explosions were, as the main one, accompanied by bright flashes, but their relatively weak flashes could not have burnt the Tunguska vegetation. The vast burn of the vegetation in the Great Hollow was only caused by the light emission from the main explosion.

And what about other Tunguska traces? The local geomagnetic storm testifies that the Tunguska explosion was accompanied by ionizing radiation. At this point this is the only interpretation of the effect that is justified and substantiated by mathematical calculations. The genetic mutations of plants, insects, and humans, as well as the anomaly of thermoluminescence, do back up this conclusion.

The presence of feeble but noticeable radioactive fallout after the Tunguska explosion is another empirical fact, confirmed by finding the peaks of radioactivity dated 1908 in trees that had withered before 1945 (that is, before the year when nuclear tests in the atmosphere started and the artificial radionuclides began to fall from the sky in large numbers). Only the increased radioactivity of the samples taken from the trees that continued their growth after this year may be explained away with reference to contamination from contemporary nuclear tests.

Are "radioactive anomalies" at Tunguska weak? It depends which ones. The peaks of radioactivity in tree rings yes, but to call the thermoluminescent traces of radiation weak would not be correct. Besides, what does the expression "a weak effect" mean? It means that the effect is real; it goes beyond the limits of possible instrumental errors and therefore hypotheses pretending to account for the Tunguska phenomenon must not ignore it.

Incidentally, the most important trace of this phenomenon – the supposed material remnants of the TSB – is, as we know, also indistinct: their mass does not exceed one ton, or even several hundred kilograms. This is much too little even for an icy comet core, let alone a stony asteroid... If, say, in the Tunguska explosion 99% of the TSB substance vaporized then its mass before the explosion was just 100 tons (which is equal to the mass of the orbital stage of the space shuttle) and if it was 99.99% that disappeared then 10,000 tons (which is approximately equal to the mass of three Saturn V carrier rockets that placed the *Apollo* spacecraft in the trajectory of their flight to the Moon). The "million tons," which are frequently considered the mass of the TSB, are therefore from the realm of sheer fantasy. The real mass was considerably less.

Yet, we seem to have digressed from facts to hypothetical constructions. Let's return to reality.

To solve the Tunguska problem we need, first and foremost, to determine the chemical composition of the TSB. So, what is now known about it? Sorting out the substances that have been discovered in the Tunguska soil and peat, we can compile the following list of 12 chemical elements whose concentration at Tunguska is unusually high:

1. ytterbium,
2. lanthanum,
3. lead,
4. silver,
5. manganese,
6. zinc,
7. barium,
8. titanium,
9. copper,
10. tantalum,
11. mercury and
12. gold.

An impressive set, isn't it? It looks rather exotic. Nevertheless, the first five elements from it – ytterbium, lanthanum, lead, silver, and manganese – not only demonstrate an increased concentration in the soil and peat but the zone of their increased concentration runs directly under the TSB's trajectory. Therefore, they could have been part of this space body. And as we've seen, the accelerated growth of Tunguska vegetation, especially pines, does also testify to a considerable contribution of rare earth elements (such as ytterbium and lanthanum) to the Tunguska site. In experiments only lanthanum and ytterbium (from the elements discovered at Tunguska) could stimulate the process of sprouting of pine seeds.

So how could a space body consisting of these elements explode? Or perhaps, we are dealing here with those components of the TSB substance which did *not* explode, and the space body consisted of two different parts – an "explosive" part and a "shell"? We can see that the complicated ("butterfly-like") outlines of the area of the leveled forest tell us that the blast wave acted unevenly, its power being very different in different directions. The strongest blasts hit the "butterfly's wings."[7] Obviously, an area of forest leveled by an even blast wave would have been shaped like a circle or, for the moving source of the blast wave, an ellipse (with some nuances, caused by peculiarities of the local terrain) – but definitely not a shape like a butterfly. If the magnitudes of the blast wave and ballistic shock wave had been comparable, one could have attempted to explain this strange shape by their interaction. But as we know, the ballistic shock wave was much weaker than the blast and therefore could not have influenced it in a significant way. Rather, the butterfly could have originated as a result of the explosion of something like a shaped charge – that is, a piece of explosive inside which a conical cavity is made and coated with a layer of metal. The blast wave destroys the cover within the hollow, starting from its top and giving enormous speed to particles of the metal. Naturally, the direction in which the blast wave of such an explosion acts most destructively coincides with the axis of symmetry of the hollow in the piece of explosive.

It was in 1959 at a conference in Moscow dedicated to the results of the first postwar academic expedition to Tunguska that the Soviet specialist in the physics of explosions – Academician Mikhail Sadovsky – said that judging from the forest destruction the source of the blast wave must have had a complicated shape.[8] The Academician had profound intuition. In those years nobody could have suspected that the outlines of the Tunguska area of the leveled forest would be as unusual as to resemble a butterfly. Subsequently, the conclusion about an intricate shape of this source of the explosion was mathematically justified by Siberian scientists Dmitry Demin and Victor Zhuravlev.

Well, let's agree that the TSB could incorporate, figuratively speaking, an "explosive" and, less figuratively speaking, a "shell." And inside the shell there were some hollows where explosions took place. But what can we say about *properties* of this "explosive"?

Attempting to explain the Tunguska explosion, authors of various hypotheses have used almost all known types of explosions: physical (impact, thermal, and dynamical, such as the swift fragmentation of the meteor body); chemical, including the vapor cloud explosion; and nuclear (fusion, fission, and antimatter annihilation). But the nuclear explosion differs very much from the chemical and physical – and not only by its magnitude. Having piled in one place 50 million tons of a powerful chemical explosive in bars and blown them up, we would not obtain all the effects that accompany the explosion of a 50-Mt thermonuclear charge. The point is that the nuclear explosion differs from all other types of explosion by its much greater concentration of energy. One cubic centimeter or one gram of a "nuclear explosive" produces *20 million times* more energy than an equivalent volume or mass of any other explosive and *100,000* more energy than is released when a meteorite collides with Earth's surface flying at a great cosmic velocity. (Let's recall, however, that there was at Tunguska no collision with Earth's surface.) Thus, according to the concentration of energy, all explosions may be separated into two groups: nuclear (having a high concentration of energy) and nonnuclear (having a low concentration of energy).[9] And what can we say about the concentration of energy of the Tunguska explosion?

Dr. Victor Zhuravlev has been studying this question in detail and for a long time and has examined the "Anfinogenov's butterfly," that is, the zone of complete destruction of the taiga. This is distinct from the larger Fast's butterfly having an area of 2,150 km^2. The area of the "Anfinogenov's butterfly" is "just" 500 km^2 – less than one-fourth of the latter. (Generally speaking, this is not so small an area. If the zone of complete destruction had looked like a circle, its diameter would have been as large as 25 km.)

It is from the area of the "Anfinogenov's butterfly" that one can calculate, using formulas from the theory of *nuclear* explosions, that the magnitude of the Tunguska explosion was in the range between 40 and 50 Mt. However, if one is using in one's calculations the equally exact formulas from the theory of *chemical* explosions, the magnitude of the Tunguska explosion turns out, strangely enough, much higher – up to 150 Mt. Why such a difference between the "nuclear" and "chemical" estimations? After all, when the Russian specialist in powerful explosions – Professor Ivan Pasechnik – used a calculation method that does not depend on the nature of the explosion (the analysis of Tunguska seismograms), he concluded that the "nuclear" figure was correct and the most probable magnitude of the explosion was 40–50 Mt.[10] The cause of the divergence lies in the essentially different levels of concentration of energy of these two types of explosion. So whatever the nature of the Tunguska explosion, its concentration of energy exceeded that of conventional explosions by about *10 million times*.

The doubts about the chemical or kinetic (impact) nature of the Tunguska explosion lie in the calculations of Alexey Zolotov when preparing his dissertation. Zolotov was reasoning from probably the most precise and informative data, namely the barographic records made in Russia and in Britain immediately after the Tugnuska explosion. Before World War II, when methods of analysis of barographic disturbances generated by powerful explosions were still underdeveloped, attempts of Francis Whipple and Igor Astapovich to use British and Russian barograms to determine the Tunguska explosion magnitude led to too low figures (maximum 50 kt, or about "four Hiroshimas"). But soon after the end of the war, the British meteorologist R. S. Scorer conducted the first professional examination of these data. And his result was 90 Mt.[11] Today 40–50 Mt is considered a more realistic figure, but the order of magnitude has remained the same. Thus, we should give Scorer his due – the more so that he had no idea about the area of the flattened forest and the number of leveled trees and therefore could not use this information in his calculations. Scorer's computations were based exclusively on the barographic data.

These barographs[12] did not record the sound waves that we hear but the so-called infrasonic acoustic waves, whose frequency is lower than we could hear. Sound waves fade very quickly in the atmosphere so that sound generated even by a very powerful thermonuclear explosion can be heard not farther than a few hundred kilometers from its epicenter. As distinct from this, infrasonic waves of such an explosion may encircle the globe several times, being recorded each time on the tapes of sensitive instruments. It was well understood as far back as 1963, when the partial Nuclear Test Ban Treaty was drawn up and signed, that characteristics of these waves might be measured at great distances. What is more, if we have barographs at several points we can determine the place and time of the explosion, as well as its magnitude. But initially, it remained unclear if it would be possible to differentiate nuclear explosions from other types of explosions – say, volcanic and conventional chemical explosions. Russian geophysicists, Professor Leonid Brekhovskikh and Professor Ivan Pasechnik, successfully solved this task, proving that "signatures" of nuclear and nonnuclear explosions on barograms are radically different.

The most evident difference between them lies in the shape of the line they trace out on the barogram. The barogram of an explosion having a low ("non-nuclear") concentration of energy looks like a wave whose size and period remain practically constant. However far the barograph is from the epicenter, its recorded timings will always be the same. If a record, made at a hundred kilometers from the epicenter, lasts 10 min, one can be sure that at a distance of 5,000 km it will also last 10 min. Yet, for an explosion with a high ("nuclear") concentration of energy the curve on the tape of a barograph will be absolutely different. We can see (see Figure 11.1) that with time both the amplitude and the period of this wave swiftly diminish. And, as distinct from a conventional explosion, the farther the barograph is located from the epicenter of the nuclear explosion, the longer will last the recording itself (from several minutes at a distance of several hundred kilometers to half an hour at several thousand kilometers).[13] It is thanks to these characteristics of air waves that specialists monitoring the observance of the Treaty of 1963 can say immediately, not waiting for information about nuclear contamination of the atmosphere, whether a powerful explosion detected by their instruments at a far-off island somewhere in the Pacific was nuclear or not.

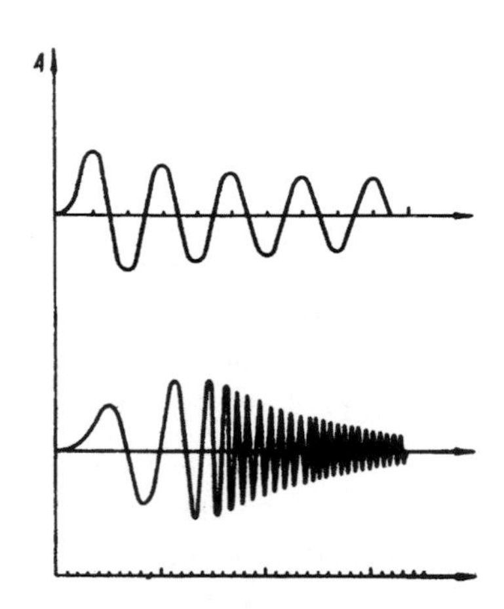

FIGURE 11.1. Here are idealized barograms of a nuclear explosion (bottom) and a nonnuclear explosion (top) compared. One can see that they are dissimilar (*Source*: Zolotov, A.V. *The Problem of the Tunguska Catastrophe of 1908*. Minsk: Nauka i Tekhnika, 1969, p. 150.).

Let's look at Figure 11.2, where barograms of a powerful chemical explosion are represented and a nuclear explosion with magnitude of several megatons that was carried out at a US testing

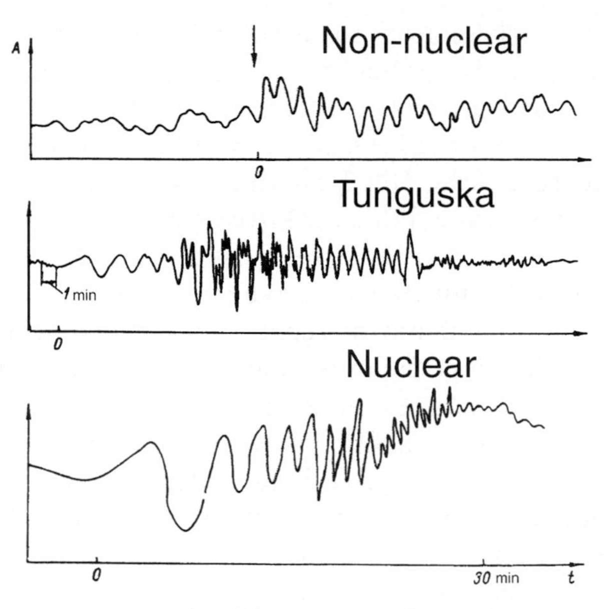

FIGURE 11.2. Comparison of real barograms of a nuclear, non-nuclear and Tunguska explosions. The Tunguska barogram does resemble the nuclear one, being very different from the non-nuclear barogram (*Source*: Zolotov, A.V. *The Problem of the Tunguska Catastrophe of 1908*. Minsk: Nauka i Tekhnika, 1969, p. 150.).

ground on Marshall Islands in 1954. A third curve is a record of air waves from the Tunguska explosion. The recording was made in 1908 in London (South Kensington). One can see that the "Tunguska" curve is very similar to the "nuclear" one, bearing at the same time no resemblance to the "chemical" curve. As for the periods during which the Tunguska barograms were recorded, in Kirensk (at a distance of 490 km from the epicenter) it was some 3 min; in Pavlovsk (3,740 km) 20 min; and in London (5,740 km) 35 min. If the concentration of energy of the Tunguska explosion had been much lower than the "nuclear," the durations of these records would have been equal. So, after comparing these curves and figures, Alexey Zolotov did have the right to say: "The explosion of the Tunguska space body had a very high concentration of energy in a small volume."[14] Somewhat later, he even took a risk to estimate the mass of this "high-concentrated explosive" that had to react in the Tunguska explosion. His final figure was just about half a ton.

Thus, information from the barograms has made it possible to establish – empirically and not referring to any hypothesis – a very important characteristic of the "TSB explosive" – its high concentration of energy. In turn, this fact confirms the conclusion about the complex structure of the TSB. It should have consisted of an "explosive" and a "shell" around this explosive; otherwise, its whole mass would have been too low to leave in the flattened forest even that weak trace of the ballistic shock wave that it did leave.

On the other hand, having agreed not only with the very great magnitude of the Tunguska explosion but also with a high concentration of its energy, which hints at the high temperature of the fiery ball, we find ourselves facing a new problem: how would it be possible to explain the herringbone pattern that exists, as we know, in the *western* part of the leveled forest area? This pattern testifies that a fairly massive body flew westward *after* the explosion. But for an explosion with a near to nuclear concentration of energy, according to the barographic data, the TSB's survival looks incredible. In fact, no material body could have survived this hellfire. If something did in fact pull through, this means there were *two* space bodies, one of which had exploded and another that continued flying to the west. (This idea about two bodies, by the way, follows from the existence of two compact groups of eyewitnesses – the southern and eastern ones – as well as from two axes of symmetry of the butterfly-like area of the leveled forest, determined by Wilhelm Fast.)

Now, we have outlined 25 components of an interdisciplinary model of the Tunguska phenomenon – from the low velocity of the TSB's motion and its peculiar chemical composition to a high concentration of energy in the Tunguska explosion and its directional character – using for this conclusion the 10 Tunguska traces, records of barographs and seismometers, plus the eyewitness testimonies. So which of these parameters of the Tunguska phenomenon are more reliable and which are less reliable? The most reliable parameters are, naturally enough, those that have been reflected in *several* traces. However strange it may seem, these are those features of the phenomenon that look very unusual from the viewpoint of traditional cometary and asteroidal hypotheses – for a start, the presence of ionizing radiation. There are four traces pointing at this: the local geomagnetic storm, genetic mutations, anomalies of thermoluminescence, and radiation peaks in the trees that had withered before 1945. Also, such an unexpected fact as the participation of *two* space bodies in the Tunguska event may be derived from three traces: two separate groups of eyewitnesses – in the south and in the east, two axes of symmetry of "Fast's butterfly," and the observation by shaman Aksenov of a flying body after the explosion.

And so, having at our disposal all these data, let us look at what can be concluded. Beginning on the evening of June 27, 1908, some space body was orbiting Earth, and by its motion disturbing the geomagnetic field. These magnetic disturbances were recorded in the German city of Kiel by Professor L. Weber. Also, in the same days in some places of western Europe, observers reported atmospheric optical anomalies. Soon after midnight GMT on June 30, 1908, just while the Weber effect was being recorded for the last time, two space bodies, flying at a relatively low speed, entered the atmosphere of our planet. They passed over central Siberia, moving toward the Great Hollow, the slopes of their trajectories not exceeding 15°. One of these bodies – let's call it TSB-A – flew from the south to the north, and the second – TSB-B – from the east-southeast to the west-northwest (see Figure 11.3).

The "southern" TSB-A flew over the Angara River not far from the village of Kezhma, flying more or less in a straight course (at least, we have no information about any maneuvers performed by it). The "eastern" TSB-B first traversed the upper reaches of the Lena River near the village of Mironovo and then the upper reaches of the Lower Tunguska River over the village of Preobrazhenka, flying in an arc. Having approached the Great Hollow and flying at several dozens of kilometers to the north from Vanavara, both the bodies changed direction. The TSB-A turned to the west-northwest and the TSB-B almost to the west. At an altitude of 6–8 km, there occurred an explosion annihilating the TSB-A, leveling 30 million trees, burning by a light flash an area of more than 200 km^2, and producing a forest fire. The explosion had been uneven and very powerful – comparable in its magnitude with the explosion of the "Tsar-bomb" that was tested on the Soviet nuclear testing ground Novaya Zemlya in 1961.

The TSB-A exploded due to an inner energy, not due to kinetic energy, its concentration exceeding considerably the level that is possible for conventional explosives and approaching that of a

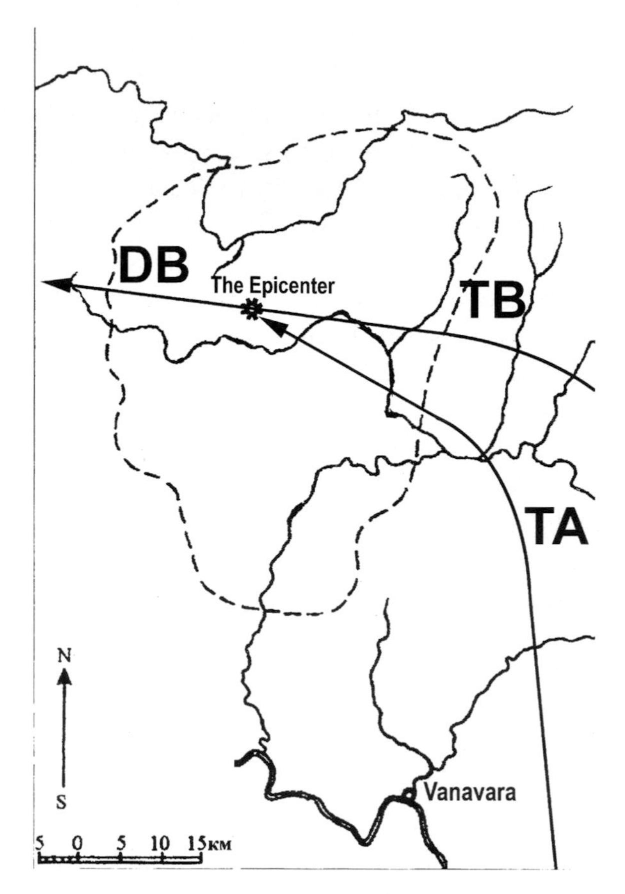

FIGURE 11.3. Directions of approach of the first (TA) and the second (TB) Tunguska space bodies to the epicenter; the trajectory of departure of the surviving body (DB).

nuclear explosion. But, most probably, only approaching but not reaching that level, as evidenced by the fact that separate pieces of the TSB-A were still exploding during a couple of minutes at lesser altitudes and with considerably less power. It is quite obvious that a nuclear charge would not have left any pieces after its explosion. The fiery ball, formed during this explosion, rose to the upper atmosphere, where its ionizing radiation induced a magnetic disturbance in the ionosphere. It developed into a local geomagnetic storm lasting about five hours. Products of the explosion (which contained, judging from the data of the Mount Wilson Observatory, some peculiar aerosol of ultramicroscopic particles suspended in the air) got into the atmosphere and gave rise to an abrupt jump in intensity of anomalous atmospheric phenomena over western Europe and European Russia.

Immediately before the explosion, the TSB-A was flying relatively slow (at a velocity not exceeding a couple of kilometers per second), its diameter being about 50 m. It appears that the structure of the TSB-A was far from uniform, due to which the blast wave acted most strongly in two directions – to the south-southeast and east-northeast, forming the butterfly-like shape of the area of the leveled forest. Only a very small share of its mass (some five thousandth of a percent) had reacted in the explosion, its whole mass being not more than 10,000 tons. The lack of a long tail of burning substance behind this body, when it was moving through the atmosphere, indicates that it did not lose any noticeable mass due to ablation – that is, the loss of surface material through evaporation caused by friction with the atmosphere. The TSB-A had a fairly low average density, but sufficiently high mechanical strength. And the paleomagnetic anomaly, discovered in the Great Hollow, testifies that it was also a source of a powerful magnetic field.

The TSB-B continued its flight westward – possibly gaining altitude (otherwise it would have fallen not far from the epicenter and flattened the taiga even more). Nothing concrete is known about its physical parameters (dimensions, mass, velocity at this stage of flight), but since the "herringbone" trace left by it in the western part of the area of the leveled forest was weaker than a similar structure left by the TSB-A in the eastern part of this area, its mass and/or velocity must have been less than those of TSB-A. For good reason, we know absolutely nothing about its chemical composition. But as for the chemical composition of the TSB-A, the main 12 elements of which it could consist were listed earlier – from ytterbium to gold.

Just 15 min after the Tunguska explosion, the Weber effect stopped and it never returned. Probably, the space body that had been producing it left near-Earth space (whether "upward" or "downward").

It is worth noting that the above description of the Tunguska phenomenon does not pretend to be exhaustively complete or absolutely accurate. Quite possibly it lacks some important details (just because these had not impressed themselves in the Tunguska forest, soil, and peat, or on the bands of seismographs and barographs) or that some characteristics have been represented imperfectly. But this model has one very important advantage over all other "theoretical" pictures of the Tunguska event: it has been built on the real empirical facts, any hypothetical consideration having been ignored. Certainly, the proposed picture is open to change and criticism. But it would be desirable to have these changes and this criticism also based on facts and not on preferred theories.

One must admit with some regret that the reconstructed image of the Tunguska phenomenon does not offer a definite answer to the question "what was it?" What is more, none of the hypotheses considered in the previous chapter – even the hypothesis by Henrik Nikolsky and his colleagues about the "orbital comet" – fits this image sufficiently well. In particular, the high concentration of energy of the Tunguska explosion contradicts the hypothesis of the vapor cloud explosion. And an ordinary comet or a stony asteroid seems to be out of the question.

Hence, the Tunguska mystery has once again demonstrated the high level of its intricacy. This does not mean, of course, that none of the existing hypotheses can be improved to convincingly explain this picture. But one should not put the cart before the horse and ignore facts just because they contradict this or that theory. The ultimate objective of science is scientific truth, however stiltedly or banally it might sound. And this objective can be reached only if the scientist is constantly comparing results of his or her abstract thinking with empirical facts. Even if it will be needed to add complexity to an existing theoretical scheme or to build a principally new theoretical scheme to account for the event that occurred in central Siberia in the summer of 1908 – well, such things have happened in the history of science. After all, we are very lucky that the set of Tunguska data, accumulated by several generations of researchers, is very detailed and informative.

It only remains to understand the meaning of these facts, details, and figures. As Albert Einstein used to say, "God may be subtle, but He isn't plain mean." Similarly, the Tunguska phenomenon is by no means trying to mislead us, but a considerable level of subtlety in it can also be noticed. It is therefore necessary for scientists to display an equally high level of ingenuity – and then the peculiar, enigmatic, and sometimes challenging facts will turn out obvious elements of a well-balanced picture.

Notes and References

1. If we accept that the TSB started to emit light at an altitude of 150 km (which may be considered as overstating for usual meteors, but admissible), then at a distance of 1,000 km from the epicenter it could be seen if the slope of its trajectory did not exceed 5°. But taking into account various additional factors (such as the radius of the field of vision of the eyewitnesses), this figure should be somewhat increased.
2. See Dmitriev, A. N., and Zhuravlev, V. K. *The Tunguska Phenomenon of 1908 as a Kind of Cosmic Connections Between the Sun and the Earth.* Novosibirsk: IGIG SO AN SSSR, 1984, p. 34 (in Russian).
3. For lovers of exact figures: the Tunguska space body exploded at 0 h 13 min 35 s GMT $\pm$ 5 s. See Pasechnik, I. P. Refinement of the moment of explosion of the Tunguska meteorite from the seismic data. – *Cosmic Matter and the Earth*. Novosibirsk: Nauka, 1986, p. 66 (in Russian).
4. Recently, there appeared a different estimation – a few megatons. We will consider this figure in the final chapter of the book.
5. See Demin, D. V., and Simonov, S. A. New results of processing the catalog of Tunguska leveled trees. – *Tungussky Vestnik*, 1996, No. 3 (in Russian); Demin, D. V. On some peculiarities of the energy-generating zone of the Tunguska phenomenon of 1908. – *RIAP Bulletin*, 2000, Vol. 6, No. 1; Golenetsky, S. P., Stepanok, V. V. Comet substance on the Earth (some results of investigations of the Tunguska cosmochemical anomaly). – *Meteoritic and Meteor Studies*. Novosibirsk: Nauka, 1983 (in Russian).
6. Suslov, I. M. Questioning witnesses in 1926 about the Tunguska catastrophe. – *RIAP Bulletin*, 2006, Vol. 10, No. 2, pp. 18–19.
7. Which also testifies that the blast wave could not originate due to the swift fragmentation of the space body – otherwise we would have seen the maximal destructions in a forward direction. See Kuvshinnikov, V. M. On some peculiarities of the Tunguska area of leveled forest. – *The Tunguska Phenomenon: Multifariousness of the Problem*. Novosibirsk: Agros, 2008, p. 161 (in Russian).
8. See Tsikulin, M. A. Shock waves generated by the atmospheric motion of large meteorite bodies. Moscow: Nauka, 1968, p. 5 (in Russian).
9. Around 4.2×10^{10} and 8.4×10^{17} ergs per gram, accordingly.

10. Pasechnik, I. P. Estimation of parameters of the Tunguska meteorite explosion from seismic and microbarographic data. – *Cosmic Matter on the Earth*. Novosibirsk: Nauka, 1976 (in Russian).
11. See Scorer, R. S. The dispersion of a pressure pulse in the atmosphere. – *Proceedings of the Royal Society of London*. Series A, Mathematical and Physical Sciences, 1950, Vol. 201, No. 1064.
12. Strictly speaking, *micro*barographs, which can measure and record very small changes in atmospheric pressure.
13. For details see Pasechnik, I. P. Science has proved that nuclear explosions can be detected anyplace. – *Priroda*, 1962, No 7 (in Russian).
14. Zolotov, A. V. On energy concentration of the explosion of the Tunguska space body. – *Zhurnal Tekhnicheskoy Fiziki*, 1967, Vol. XXXVII, No. 11, p. 2094 (in Russian).

12. So What is the Answer?

Early morning on June 30, 2008, two helicopters appeared over the Great Hollow. The weather was excellent, the same as it had been a century ago – a perfectly clear blue sky, bright Sun, and heat above 30°C. Through the open portholes the fresh wind of Tunguska was blowing into the passenger compartments of the helicopters. The flying machines had arrived from Vanavara (Figure 12.1), having aboard participants of the centenary Tunguska conference, as well as TV journalists.

FIGURE 12.1. Vanavara, the closest settlement to the place of the Tunguska explosion, and the Podkamennaya Tunguska river. View from a helicopter (*Photo by* Vladimir Rubtsov.).

The 100th anniversary of the enigmatic event was a good pretext to inform the public about lots of facts and strange rumors. Just the previous evening the announcer of Central Russian Television, a very beautiful lady, informed her audience that a 100 years ago an enormous meteorite had fallen in the Siberian taiga, producing a crater 1 km across. Even if, when visiting the site, TV people have

V. Rubtsov, *The Tunguska Mystery*, Astronomers' Universe,

DOI 10.1007/978-0-387-76574-7_12,

somewhat brushed up on their knowledge of this event and know that there is no crater at all, yesterday's information had already found its way to the minds of many millions of Russian TV viewers.

The group I was with was flying more or less on the trajectory of the TSB-A, although considerably lower – at an altitude of some 800 meters. If a century ago somebody could have looked out of the Tunguska space body, he or she would have observed what we were seeing: the infinite green ocean of taiga, lakes, rivers, and no sign of humanity. This was the National Nature Reserve *Tungussky*, established in 1996 by the Federal Government of Russia, occupying an area of 3,000 km^2 and kept in its primordial state. But our impressions of this wild landscape were somewhat alarming because it seemed that time had moved backward and any moment we would see in the sky a space body performing its enigmatic maneuvers.

The helicopters first landed near the famous Kulik's Pier at the Khushmo River where, in 1927, Leonid Kulik had gone ashore from a raft and helped down the expedition's horse – the only land transport of the travelers and their last food reserve. (See Figures 12.2 and 12.3.) It was probably not easy for the horse to clamber onto the steep bank of the Khushmo – and the hordes of bloodsucking insects

FIGURE 12.2. Kulik's Pier at the Khushmo River, the place where, in 1927, Kulik's expedition debarked from its rafts. View from a helicopter (*Photo by* Vladimir Rubtsov.).

FIGURE 12.3. Vladimir Rubtsov, author of this book, at the Khushmo River.

must have been an added discomfort. More than 80 years later, similar hordes attacked us momentarily and furiously. Both journalists and scientists immediately started to button up their coats tightly and douse themselves with insect repellents. We don't know whether the meteorite hunters of the 1920s had any such repellents, or, if they did, how effective they were. Siberian bloodsuckers have never been mentioned in Kulik's publications – probably as a trifle not worth attention.

But although fighting with spiteful insects, we were already standing on the Tunguska ground, where the vast area of leveled forest was on the verge of disappearance (see Figure 12.4), but the distinct sensation of mystery still persisted. It was that very place, where a hundred years ago occurred the highly enigmatic event known as "the fall of the Tunguska meteorite." Since then its mystery has been disturbing the peace of many minds.

From the pier we headed to Kulik's *zaimka*, which is almost at the epicenter of the Tunguska explosion. On the occasion of the centenary, the authorities of Evenkya have allocated the necessary funds for the restoration of the log cabins and *labazes* (storehouses on poles) that had been built by participants of his expeditions. Near one of these has been erected a memorial sign, resembling an obelisk in honor of a crashed spaceship rather than a simple marker

FIGURE 12.4. Remains of trees, uprooted in 1908 by the blast wave of the Tunguska explosion that can still be seen in the taiga (*Credit*: Vitaly Romeyko, Moscow, Russia.).

indicating the place of a meteorite fall (see Figure 12.5). Here everyone returned to his or her own duties: journalists started to video record the landscape, and members of the international Holocene Impact Working Group studying recent impacts in the history of our planet began preparing their expedition through the surroundings of the epicenter. Several other people went on a tourist trip by the Tunguska rivers. Generally, in Russian state nature reserves, visits by tourists are forbidden, but the Tunguska nature reserve is exempted from this rule, and everyone wishing to visit this area with its unforgettable aura can do so. By the way, the first foreign tourist at Tunguska was in 1989, when the Japanese scientist Professor Kozo Kovai, a specialist in electronics, visited the region. For some strange reason he believed that in 1908 there had exploded in Siberia a spaceship piloted by a Japanese crew – and he performed a commemoration service at the site.

Yet certainly, the centenary of the Tunguska explosion has given occasion not only for excursions to Tunguska but also to more than half a dozen scientific conferences on this subject that were held in Moscow, St. Petersburg, Tomsk, and Krasnoyarsk. In Moscow two conferences were organized and three were organized in Krasnoyarsk – which is, of all large Russian cities, the nearest to

FIGURE 12.5. A memorial sign erected in honor of the centenary of the Tunguska explosion at Kulik's *zaimka*, not far from the epicenter.

the site, Evenkya being one of the administrative districts of the Krasnoyarsk Territory. Despite the area of this district exceeding that of Ukraine or Texas,[1] there live here today just 20,000 people who completely lack permanent roads, let alone railway lines. The main transport here is airplanes and helicopters, and sometimes riverboats.

The centenary of the Tunguska event, plus the 50th anniversary of the Independent Tunguska Exploration Group (ITEG) – the leading scientific research body engaged in Tunguska investigations – was a good opportunity to look back and estimate future prospects for the problem. It so happened that this author only attended the Krasnoyarsk conferences – this being the optimal choice. Pity, however, that I could not go to Tomsk – participants of that conference, as it appears to me, approached the Tunguska problem most responsibly. The conference was organized by the ITEG and its resolution states very frankly: "The Tunguska problem has not been solved as yet." At other conferences some researchers were of a different opinion, being sure that the TSB was "definitely either a comet core or a stony asteroid." Of course, they have the right to think so. But judging from a great number of mutually contradicting hypotheses that were discussed at these meetings, the Tunguska problem is still far from having been solved.[2] Its history has not been brought to an end as yet. This is a history in progress. To understand this, it is sufficient to compare what scientists had known about the Tunguska event after Kulik's expeditions of the 1920s–1930s and what they know about it today, after the ITEG expeditions of the 1960s–2000s. But this progress does have its origins in Kulik's works, and participants at the Krasnoyarsk conference felt the winds of history when the chair was taken by a daughter of Leonid Kulik – Dr. Irina Kulik, who spoke about investigations that had been carried out by her father. Sir Isaac Newton once said briefly and wisely: "If I have seen further it is by standing on the shoulders of giants." Tunguska investigators of the twenty-first century also see further than previous generations of researchers for the same reason. This is very important in science.

True, the Tunguska centenary also gave rise to new "jubilee solutions" of the problem. In the former Soviet Union the authorities liked to have politically important events and anniversaries marked by bright scientific and technological achievements that had a broad effect – sometimes all over the world. For example, the second *Sputnik* with the dog Layka aboard was launched by personal command of Nikita Khrushchev to celebrate the 40th anniversary of the October Revolution; the "Tsar-bomb" with its 50 Mt of explosive power was tested to mark the 22nd Congress of the Communist Party of the Soviet Union, and so on. So scientists and engineers participating in such projects had more chances for high governmental awards than those involved in ordinary "non-jubilee" events.

Maybe this sort of thing seldom happens in the western world, but the desire to celebrate the centenary of a problem by its solution sometimes takes place – and why not? So, Dr. M. B. E. Boslough and Dr. D. A. Crawford at the Sandia National Laboratories in the United

States provided a gift for Tunguska's 100th birthday. They devised and simulated on the world's fastest supercomputers an innovative mathematical model of the Tunguska event.

These researchers took as a basis for their computations the results of observations of the comet Shoemaker-Levy's fall on Jupiter in 1994, when an upward-directed atmospheric plume in the atmosphere of that planet had been detected, as well as the assumption that Tunguska cannot be treated as an isotropic explosion. Instead, according to Boslough and Crawford's theory, "the wake of the entry creates a low-density, high-pressure channel from the point of maximum energy all the way out of the atmosphere, so the explosion is directed upward and outward."[3]

Naturally enough, under this assumption the whole magnitude of the Tunguska explosion must have been much less than if its energy propagated evenly in all directions (Boslough and Crawford's calculations have led them to the figure of some 3.5 Mt). This result was strained: the authors stated that the terrain around the Tunguska epicenter looks like a slope of 15°. This is not so: there are there slopes directed both from the epicenter and toward it. However, this is not too important. Let the magnitude of the explosion be somewhat more than 3.5 Mt – say 5 or even 7 Mt. But which parameters of the TSB's motion have been used in the Sandia model? Alas, purely "theoretical" ones: a stony asteroid having a mass of some 350,000 tons had been flying at a velocity of 15 km/s at an angle of 45° to Earth's surface. The reader does certainly understand that this angle sharply contradicts the reliable testimonies of eyewitnesses of the Tunguska phenomenon. It is easy to calculate that flying in such a trajectory at a distance of 1,000 km from the epicenter, where the TSB was already brightly visible, its altitude would have been 1,300 km. A material space body could emit light at this altitude, in a complete vacuum, only if somebody had placed on it festive illuminations.

Also, the Sandia specialists are completely silent about the shape and structure of the area of the leveled forest after their computed airburst – promising to accomplish, with time, "a full 3-D simulation of various Tunguska scenarios using a high resolution model of the actual topography of the site." When and if such a future simulation shows something resembling the "forest butterflies" of Fast's and Anfinogenov's, the Sandia model will be worth further discussion. But for the time being it remains just another mathematical construction having a very distant relation to the Tunguska event as such. To attack the Tunguska problem, ignoring characteristics of the area of the leveled forest is the same as computing parameters of the Arizona meteorite, having no idea of the shape, dimensions, and depth of the crater it has left. Would the results of such a computation have had anything to do with the real event that had occurred in the Arizona desert some 50,000 years ago? Hardly so.

Let us add that not a single Tunguska eyewitness saw in the sky any plume – which, according to the Sandia scientists, must have been ejected backward along the TSB trajectory. Such a plume would certainly have been noticed. As for the attempts of Boslough and Crawford to use the alleged plume for the explanation of the after-catastrophe illumination in European skies, these are simply absurd: the ejected TSB substance must have dispersed in the atmosphere to the east from the place of the explosion from which the space body arrived; but Europe is located *to the west* – in the opposite direction. Finally, we must ask the same time-honored question: where is the substance of their "Tunguska asteroid," those thousands or even tens of thousands of tons of rock that had to be scattered over the Great Hollow? The Sandia scientists are referring to the work of Moscow physicist Dr. Vladimir Svettsov, according to which the TSB substance had been completely vaporized by the light flash; but Svettsov's conclusion has been convincingly refuted by Vitaly Bronshten and Andrey Olkhovatov: complete vaporization of a stony asteroid is impossible, the region of the Tunguska epicenter would have been strewn with meteoritic dust and even with fairly large pieces of the "heavenly rock."[4]

One cannot but agree with Dr. Victor Zhuravlev, who wrote in 2006: "The main distinctive feature of the contemporary stage of Tunguska investigations is the wide gap between the concrete results of expeditions which crossed the Siberian taiga, were digging in Tunguska soil and peat, measuring thousands of leveled trees, questioning eyewitnesses about the phenomenon, and, on the other hand, the theoreticians who are building computer models of the phenomenon. This gap is now the main obstacle to the further development of this field of research."[5]

Those wishing to find out what did in fact happen in central Siberia in 1908 have to consider the whole body of relevant data;

only then will a realistic model of the phenomenon be seen through the apparent chaos of this body of information. In previous chapters we deliberately paid much attention to the history of the Tunguska problem: these are not just old tales having nothing to do with the current state of the problem. Instead, this is the path of the successive approximations to its solution. Having gathered together all known material and instrumental traces of the Tunguska event, as well as having analyzed eyewitness reports, we have built on this basis a multidisciplinary picture of the phenomenon – but it turned out not to correspond with the existing hypotheses.

Probably it would be worthwhile to try and computerize this picture one day in the future, using up-to-date algorithms and programs. Then we would be able to find out which empirical data are still lacking, despite many years of hard research work at Tunguska, and to start looking for it. But even now the present picture may be considered a good approximation to the truth. And, as the famous detective Sherlock Holmes used to say, "the more bizarre and grotesque an incident is the more carefully it deserves to be examined, and the very point which appears to complicate a case is, when duly considered and scientifically handled, the one which is most likely to elucidate it."

The situation in Tunguska research would have looked much more hopeless if we had had no bizarre traces – neither the rare earth anomaly, nor indications of genetic mutations, nor the very informative barograms, nor everything else. In this case, researchers trying to unravel this mystery would probably have had to seek the help of a "natural non-local explosion." Yet at present there is no need to despair: we are perhaps within a couple of steps from the final solution of the Tunguska problem.

But what may be this final solution? How must it look and how may it be achieved? Of course, even such detailed theories as those developed by Grigoryan or by Boslough and Crawford cannot be considered as solutions, much less as final solutions. These are just hypothetical models whose validity is still to be tested in the field. All experienced Tunguska specialists agree that this problem will be solved only when a real piece of the Tunguska space body has been found. One can elaborate an imposing theory of the Tunguska explosion, full of equations and mathematical functions, but the only method of its verification may be discovering appreciable quantities of the TSB substance in an area predicted by theory. Otherwise competition between various viewpoints could last forever.

Of course, it is not difficult to call for a search for a piece of the TSB – but how can it be found if the whole Tunguska enigma had largely arisen due to the lack of such an item? Leonid Kulik expended plenty of time and effort drilling the empty thermokarst holes for nonexisting pieces of the TSB. True, at present we have in Tunguska data some hints about the substance (ytterbium, first of all – and the whole list of 12 elements), but these are just hints – literally microscopic hints. That is why, when building new Tunguska hypotheses, the majority of scientists take the liberty of ignoring them.

But as a matter of fact, the only thing of which the Tunguska investigators are today certain is the lack of considerable quantities of the TSB substance, *more or less uniformly covering the Great Hollow*. Of course, nobody can guarantee that one or two fairly large pieces of the Tunguska space body are not lying somewhere in the Great Hollow, under a layer of soil or peat. They may be hidden in Lvov's bog – a peat bog near the northwestern slope of the Ostraya Mountain. As Victor Zhuravlev remarked in 1998, just here such intriguing anomalies in the Tunguska area as mutations in pines and insects are most evident, as well as an increased concentration of ytterbium in the soil. In the 1920s, some Evenk people even recalled that after the Tunguska catastrophe they had discovered in this place some "pieces of metal, lighter in color than a knife's blade."

Dr. Zhuravlev has worked out a special research program called "Lanthanum," aimed at the search for geochemical anomalies in vertical columns of soil on the beach of Lvov's bog. The goal of this program is the gradual detection of the center of various anomalies in this part of the Great Hollow – like in geological prospecting an ore body is detected by mapping geochemical peculiarities around it. When carrying out this program, the precise coordinates of the zone of the probable fall will be determined. And a relatively large body itself might be detected, according to Dr. Vladimir Alekseev, who is also participating in this program, with the help of a new powerful georadar, made at the Moscow Institute of Terrestrial Magnetism, Ionosphere and Radio Wave Propagation of the Russian Academy of Sciences (IZMIRAN). This device makes it possible to study soil and rock down to the depth of 100 meters, displaying a three-dimensional (3D) picture of underground objects. Dr. Alekseev believes that some pieces of the TSB could have penetrated the Tunguska soil, forming no craters. Although such investigations are still in their infancy, they look promising. And if the scientists happen to be lucky and find the necessary funding, we may witness a return to the search for large pieces of the Tunguska space body – searches similar to those that had been pursued in the 1920s by Leonid Kulik. In the history of scientific investigations such returns sometimes occur.

Of course, it is necessary to study more the traces of radiation at Tunguska. During one of my last meetings with Professor Nikolay Vasilyev we discussed this direction of investigation in detail. According to Vasilyev, in the history of Tunguska investigations there existed a strange trend: attempts to find traces of radioactivity were made more than once and by various methods, but each time, as soon as a positive answer to this question began to turn up, the work was immediately interrupted. Researchers either stopped their work by their own initiative, blankly attributing the positive result to "chance contaminations," or the lack of money and technical means prevented the development of further work. In some cases, the researchers died. (Here we must emphasize that nothing suspicious was ever found in such cases, and there is no reason to fantasize about any conspiracies. Simple coincidences – but sad ones.)

Nikolay Vasilyev was pinning his hopes for further progress in the search for Tunguska radioactive trace on the thermoluminescent investigations that were carried out for a long time by Boris Bidyukov, Mikhail Korovkin, and other ITEG members. It is the thermoluminescent method that allows the detection of weak and old traces of radiation; other measuring techniques are too rough for that. It seems his hopes were not groundless. In particular, the "Deer-stone," an unusually large stone (photograph on Fig. 6.7) discovered by John Anfinogenov on Stoykovich Mountain, near the epicenter, although not a piece of the TSB (as John himself would have liked), does let us know something essential about the Tunguska phenomenon.

"Quartz samples, taken from a near-surface layer of the Deer-stone, are remarkable for the high intensity of their thermoluminescence, which is weakening as the depth of the sampling increases," wrote Korovkin and his colleagues when reporting their experiments. "We can make a justified assumption that the Tunguska explosion was accompanied by hard radiation."[6] Having made this discovery, Mikhail Korovkin ceased his research work in this field. It appears that the trend, noticed by Professor Vasilyev, still remains in force.

Luckily, not all Tunguska researchers are yielding to it. Boris Bidyukov, who assumed responsibility for the thermoluminescent investigations in ITEG in the mid-1970s, is continuing his work on thermoluminescence. His team that collected samples at Tunguska consisted of 80 people, and this work lasted several decades. In 1988, Boris decided to publish their empirical data, not trying to explain it. But in a recent Tunguska collection of papers he said bluntly: "Formerly we were calling the factor which had stimulated thermoluminescence at Tunguska somewhat too cautiously 'unknown,' but now it's time to tell that we cannot see any rational alternatives to identifying this with hard radiation."[7]

Perhaps 99% of Western scientists and science amateurs interested in the Tunguska problem, if they happen to read this statement of Boris Bidyukov, would exclaim: that's impossible! It's common knowledge that the myth about the Tunguska radiation was rebutted by somebody somewhere sometime – wasn't it? And stating this, the same people will not fail to complain that "the Russians" are inclined to consider the Tunguska problem as something close to their private property. But indeed, the members of the Tunguska research community in Russia, Ukraine, and other CIS countries, although far from uniform in their viewpoints on the phenomenon and not too diplomatic when arguing about it, do have a grasp of the real contents of this problem, whereas their Western colleagues are as a rule dealing with its simplified and perhaps distorted pictures. Too many well-established facts have been forgotten, too much information is ignored, lots of important publications remain unknown in the West – partly because of the language barrier. Besides, scientific overspecialization, so typical in this day and age, hampers the interdisciplinary perception of the Tunguska phenomenon. At best, the researcher knows that there is in Siberia an area of leveled forest, having at the same time no idea of other Tunguska traces – both larger (the light burn and the geomagnetic storm) and smaller (from genetic mutations to the paleomagnetic anomaly) or of other "details" of this event.

One should also take into account the fact that a considerable part of the empirical information, collected by ITEG people at Tunguska, has not as yet been processed. Since 1995, members of the ITEG have been discussing the idea of creating a full electronic database on the Tunguska phenomenon, but only some preparatory steps have been taken. It is evident, however, that this database will be enormous. If we simply cast a glance at the data presented in previous chapters, we can see how astonishing it is that we already know so much about the Tunguska phenomenon, and what a great number of various hypotheses have been put forward to account for it, and how many people of splendid intellect have pondered over this enigma – and yet how poorly, despite all of this, we understand its origin and nature.

So why is this? Why has such a rich set of empirical information not yet been transformed into an accurate and rational theoretical scheme explaining this phenomenon? Do we lack additional data – or something else? In fact, we can have a deep insight into the nature of the Tunguska event only due to a creative imagination – and the main trait of the creative imagination and the first condition of its effectiveness is intellectual bravery. Logic, discipline of reasoning, ability to match theoretical considerations with factual material – all these are important in the next stage of scientific investigation, the stage of testing the proposed ideas. But by hastily rejecting ideas because of their "excessive audacity," when they are only emerging, we are erecting a stony wall across the path of the progress of science, which is far from smooth even without such walls.

Perhaps then, the starship hypothesis put forward by Alexander Kazantsev in 1946, which perturbed the still water of the meteoritic pool, is not only of historical interest. Even its opponents admit with reluctance that the role of this hypothesis in the history of Tunguska investigations was very important. Just try to imagine this history without Kazantsev's idea! Meteor specialists would have never started searching for subtle traces of radioactivity, or investigating thermoluminescence, or studying genetic mutations, and all these traces of the Tunguska explosion would have sunk into oblivion. Even the shape and structure of the area of the leveled forest would have remained vague. So, it is a respectable hypothesis that greatly contributed to the development of the problem, not just a fantastic speculation. But what place does this hypothesis occupy in Tunguska studies today?

One has to admit that it went through its apex in 1969, when Alexey Zolotov published his famous monograph *The Problem of the Tunguska Catastrophe of 1908*.[8] This book has become not only the highest achievement of the "artificial" research strategy in Tunguska studies but also its swansong. Formerly the "spaceship hypothesis" had been predicting empirical facts (the overground character of the explosion, the lack of any material remnants of the TSB at the site, and so on) which then, and with much effort, supporters of "natural" conceptions were trying to explain. But from then on the situation changed. There were no new predictions resulting from this hypothesis, and the supporters of more conventional ideas had at last become able to get their breath back and to turn their attention to the details of their conceptions. Of course, the infinite waltz, performed by astronomers and meteor specialists between a comet and a stony asteroid, sometimes incorporating a carbonaceous chondrite or a cloud of cosmic dust, does not inspire the reading public, but at least nobody disturbs its performers. Unconventional but serious "natural" hypotheses (such as the "mirror asteroid" idea) do not as yet have any influence on the Tunguska problem.

There is nothing surprising in this. An interdisciplinary problem, reformulated in the language of one of the scientific disciplines that is studying it (say, in the language of ballistics), does certainly allow for a solution, acceptable to specialists in this discipline. A specialist in ballistics will write an excellent paper for a professional periodical about a particular case of motion of a large meteor body in the atmosphere of Earth. Mathematically, the problem is posed and solved on paper quite rigorously, and its solution certainly should be published. Whether or not it has anything to do with the real Tunguska phenomenon is an abstract question and academic readers will not ponder over it.

But once again, the infinite theoretical vacillations between a comet and an asteroid became possible, first of all, because the development of the "starship model" has practically ceased. Meanwhile, many specialists on the Tunguska problem believe it is far from having been refuted. In his book, which was published in 2004, Professor Nikolay Vasilyev wrote: "Calling things by their proper names without diplomatic curtsies, I would like to emphasize that of all known impact events the Tunguska phenomenon is the only one in which a contact with extraterrestrial intelligent life might be surmised."[9] And in another work: "I think you understand well: being a science professional I do realize that what I am saying is rather risky. But it must be said."[10]

Vasilyev believed that although "there are as yet no direct proofs of the contact," they "may appear if the elemental and isotopic composition of the Tunguska space body could be reconstructed." To tell the truth, here this eminent scientist was somewhat too optimistic as it seems that even the most unusual chemical composition of the space body that exploded at Tunguska in 1908 would be, in this or that way, forced into the cometary-meteoritic TSB model. And certainly, the inability of this model to account for, say, the geomagnetic storm or anomalies of thermoluminescence would not worry anyone.

How starships may be constructed we can only conjecture; but without at least a general idea of their physical principles of motion it is very difficult to interpret in terms of Kazantsev's hypothesis even the most unusual findings at Tunguska. For example, what does the paleomagnetic or rare earth anomaly tell us? In the absence of theoretical models of extraterrestrial spaceships, they only suggest that the Tunguska phenomenon could hardly have been produced by a stony asteroid or by a comet. Alas, science does not possess as yet any theoretical models of alien starships or alien artifacts. So the scarcity of "artificial" models of the Tunguska phenomenon is disappointing but understandable. The "natural" research program is in this respect much richer. But as for the "artificial" Tunguska research program, its number of hypotheses is just one. It is that an alien spaceship perished in the final stage of its flight due to a technical malfunction. However, we have to ask: is there any sense in working out different versions of this hypothesis if we cannot evaluate their plausibility?

Perhaps there is. While working in the 1970s at the "Laboratory of Anomalous Geophysics" and studying the Tunguska problem together with Alexey Zolotov, Sokrat Golenetsky, and Vitaly Stepanok, this author got accustomed to integrating empirical Tunguska data by using what could be called the "model of an aerospace combat." According to this model, there happened in 1908 over central Siberia an aerial engagement between two extraterrestrial spaceships, after which one of them survived and flew back into space. Of course, this is not meant to be offered as the final solution of the Tunguska mystery, but as a working instrument this hypothesis proved to be helpful. And the multidisciplinary model of the Tunguska phenomenon, built in the previous chapter, does not contradict it either. In the Soviet Union, however, authorities hated the idea of "star wars" and *Glavlit* (the Soviet censorship) would never have allowed the "model of an aerospace combat" to be mentioned even briefly on the pages of the scientific or popular press. This is why we did not try to propagate it, although there were discussions with Alexey Zolotov about its possible implications. To Zolotov the aerospace battle hypothesis did appear of interest, although he doubted that it could be validated on the basis of existing evidence.

The search for extraterrestrial intelligence is a legitimate field of scientific investigation. And obviously, if the Tunguska phenomenon has something to do with this, then it must attract still more attention from the science community. But paradoxically, as Nikolay Vasilyev noted in his last book, if it is not so the Tunguska problem may turn out even more important – and not only for science but for all inhabitants of this planet. Astronomers used to think that there are only two types of dangerous cosmic objects (DCOs): comets and asteroids. But if the TSB was a natural space body, then it means there exists in space another type of DCO, whose nature remains unknown. Professor Igor Astapovich, a Ukrainian scientist who contributed greatly to the Tunguska problem, wrote as far back as before World War II: "If the Tunguska meteorite had fallen 4 hours 48 minutes later then St Petersburg would have found itself in the seat of its explosion and the city would have been in ruins."[11]

Today the astronomical picture of our universe is full of catastrophes, with its Big Bang, black holes, X-ray bursters, supernova stars, and an enormous number of impact craters on the Moon, Mercury, and Mars. These don't surprise us any longer. Humanity seems to have got used to cosmic dangers, although recently the idea that impact processes could have played an essential role in the geological history of our own planet was not so well appreciated by the scientific community. The discovery in the Yucatan in 1978 by geophysicist Glen Penfield of the Chicxulub crater, 180 km across, left by a gigantic asteroid that some 65 million years ago had most likely caused the extinction of the dinosaurs, and the impressive picture of the collision of comet Shoemaker-Levy with Jupiter in 1994 have considerably weakened this negative reaction. Debates – and hot ones – are now dealing with an important issue: whether or not such collisions happened during recent human history. It would have been definitely reassuring to know that the heavens have confined themselves to the extermination of ancient reptiles and will treat mammals and humans more delicately.

However, members of the Holocene Impact Working Group, including scientists from the United States, Russia, Ireland, France, and Australia, are of the opposite opinion – that humankind is in some sense an endangered species, too.[12] According to them, gigantic tsunamis produced by large meteorites falling in oceans occur approximately every 2,000 years, destroying inhabited localities ashore and influencing thereby the course of history. This hypothesis (finding some corroborations in field investigations) has generated a squall of criticism.

Nevertheless, during the last 15 years or so science has paid some attention to potentially dangerous cosmic objects, and preliminary work for developing spaceguard systems has been carried out. This term – spaceguard – was coined by the famous science fiction writer Arthur C. Clarke, who meant by it an early warning system to detect "near earth asteroids" (NEAs) whose orbits cross the orbit of our planet. In reality, about 350 NEAs have already been detected, and scientists have found more than 200 ancient meteor craters – "star wounds" – even though they have been partly obliterated for millions of years.

In this respect, the computations of Drs. Boslough and Crawford from Sandia National Laboratories are definitely important. As they believe, "low-altitude airbursts are by far the most frequent impact events that have an effect on the ground. The next impact on Earth that causes casualties or property damage will almost certainly be a low-altitude airburst."[13] Although these considerations do not bear a direct relationship to the Tunguska phenomenon (at least not until a real 3D simulation has been made on a real 3D map and its calculated area of the flattened forest turns out to match the two Tunguska "butterflies"), their results hint that even falls of not too gigantic cosmic bodies might be fraught with grave dangers for our civilization.

Fortunately, humanity has one very useful, although sometimes thoroughly veiled quality, owing to which it survived in prehistoric times: the ability to face the truth. Let's hope it has not lost it. To hide one's head under a blanket is easy. After all, the theoretical chances for a catastrophic impact in the nearest days or months are, frankly speaking, not excessive, and conclusions of the Holocene Impact Working Group still must be confirmed by other researchers specializing in this field. But such a strategy will hardly be conducive to the further survival of humanity – if only because an "unlikely event" does not mean "an impossible event."

It would be silly to panic, repeatedly looking at the sky, waiting for a cosmic catastrophe. But it would be even sillier to forget our vulnerability on this planet. Arthur Clarke once cited a phrase of another science fiction writer, Larry Niven, with regard to the asteroidal danger: "Larry Niven summed up the situation with the phrase: 'The dinosaurs became extinct because they didn't have a space program.' And we will deserve to become extinct, if we don't have one." Sounds good, but this author would like to offer another explanation of this ancient disaster. The dinosaurs became extinct because they attempted to economize on serious investigations of Tunguska-like events that probably occurred from time to time in their Jurassic paradise.

* * *

So, dear reader, we have journeyed in this book together, through a maze of instrumental data and wild rumors, scientific hypotheses and naive inventions, and the thickets of Tunguska taiga and the near-vacuum of the terrestrial ionosphere, as well as through many other places in space and time. We hope that some Tunguska facts have become for you in this journey more understandable. Possibly, some others have become even more enigmatic.

Did we find the correct solution of the Tunguska problem when making this journey? Unfortunately not – but at least we have seen this problem in all (or almost all) its details and nuances. And a nuance is not a trifle – far from that. More often than not, the gist of a matter is hidden in its nuances. That's why it would be careless to divide them a priori into the "essentials" and "nonessentials." The Tunguska fortress has not surrendered as yet, but there are cracks in its walls and half-open doors in its towers. To enter the fortress, it only remains to make some last efforts – and the science army will win! But these efforts have to be made; nothing will happen without our effort.

Some 80 years ago there existed in the Soviet Union the so-called GIRD, the Group for Investigations of Rocket Dynamics, from which originated the Jet Propulsion Scientific Research Institute and Sergey Korolev's Designing Bureau, which launched the first *Sputnik* and Yuri Gagarin. GIRD's engineers had worked gratis, from pure enthusiasm, as scientists at the ITEG did some 30 years later and who are continuing to do so. Who knows – perhaps from the ITEG will originate a new Interdisciplinary Tunguska Scientific Research Institute, which will radically activate investigations in this field. Then it would become possible to publish a second volume of this book – in which all final answers would be given and the correct solution of the Tunguska mystery would at last be demonstrated.

But until the "scientific troops" are gathered and sent forth, the Tunguska fortress will probably continue to resist the assault of science. It is already evident that "simple" solutions, rather popular in the history of this subject, do not work. Is this strange? Not at all. Humankind is still very young and hardly completely aware of all enigmas of the world in which it lives. Many wonderful discoveries are awaiting us – perhaps just round the corner. Should we also wait for them to suddenly reveal themselves – so that the Tunguska mystery would be solved "automatically", just like quantum mechanics has made the structure of atom understandable? But such waiting may take much too long. And here let us cite an old Japanese proverb: "If you do not know what to do, take a step forward!" None of us can see what is around the corner, but we can take that first step. Take the step!

Notes and References

1. Evenkya: almost 800,000 km^2; Texas: almost 700,000 km^2; and Ukraine: a little above 600,000 km^2.
2. See *The Tunguska Phenomenon: Multifariousness of the Problem*. Novosibirsk: Agros, 2008; *The Centenary of the Tunguska Problem: New Approaches*. Moscow: Binom, 2008; *The 100th Anniversary of the Tunguska Phenomenon: Past, Present, Future*. Moscow, 2008; *The 100th Anniversary of the Tunguska Comet Body*. St Petersburg, 2008; *The Centenary of the Tunguska Meteorite Fall: A Relay Race of Generations*. Krasnoyarsk: IPK SFU, 2008 (All in Russian.).

3. Boslough, M. B. E., and Crawford, D. A. Low-altitude airbursts and the impact threat. – *Proceedings of the 2007 Hypervelocity Impact Symposium – International Journal of Impact Engineering*, in press (2007).
4. See Bronshten, V. A. *The Tunguska Meteorite: History of Investigations*. Moscow: A. D. Selyanov, 2000, pp. 223–225 (in Russian); Olkhovatov, A. Y. *The Myth About the Tunguska Meteorite. The Tunguska Event of 1908 as a Mundane Phenomenon*. Moscow: Association Ecology of the Unknown, 1997, pp. 101–102 (in Russian).
5. RB Questions and Answers: Dr. Victor Zhuravlev. – *RIAP Bulletin*, 2006, Vol. 10, No. 1.
6. Korovkin, M. V., Gerikh, L. Y., Lebedeva, N. A., and Barsky, A. M. Assessment of radiation conditions in areas of ecological instability by methods of radiation mineralogy. – *Radioactivity and Radioactive Elements in Human Environment*. Tomsk: Tomsk Polytechnic University, 1996 (in Russian).
7. Bidyukov, B. F. Thermoluminescent investigations in the region of the Tunguska catastrophe. – *The Tunguska Phenomenon: Multifariousness of the Problem*. Novosibirsk: Agros, 2008, p. 83 (in Russian).
8. Zolotov, A. V. *The Problem of the Tunguska Catastrophe of 1908*. – Minsk: Nauka i Tekhnika, 1969 (in Russian).
9. Vasilyev, N. V. *The Tunguska Meteorite: A Space Phenomenon of the Summer of 1908*. Moscow: Russkaya Panorama, 2004, pp. 12–13 (in Russian).
10. Vasilyev, N. V. Memorandum. – *Tungussky Vestnik*, 1999, No. 10 (in Russian).
11. Astapovich, I. S., Fedynsky, V. V. *Meteors*. Moscow: Gostekhizdat, 1940, p. 79 (in Russian).
12. This group includes Dr. Dallas Abbott from Lamont-Doherty Earth Observatory in Palisades, N. Y., Dr. Bruce Masse from Los Alamos National Laboratory, New Mexico, Dr. Viacheslav Gusiakov, Head of the Academic Tsunami Laboratory in Novosibirsk, Russia, and other specialists from various countries. See, for example: Masse, W. B., Weaver, R. P., Abbott, D. H., Gusiakov, V. K., Bryant, E. A. Missing in action? Evaluating the putative absence of impacts by large asteroids and comets during the Quaternary Period. In: *Proceedings of the Advanced Maui Optical and Space Surveillance Technologies Conference,* Wailea, Maui, Hawaii, 2007, pp. 701–710.
13. Boslough, M. B. E., and Crawford, D. op. cit.

Index

U

V

Printed in the United States of America